Basic and Clinical Aspects of Neuroscience Vol. 4

Edited by C. Weil (Managing Editor)
E. E. Müller and M. O. Thorner

Springer Sandoz
Advanced Texts

SANDOZ

Somatostatin

With Contributions by

Y. C. Patel J. Epelbaum D. R. Rubinow C. L. Davis
R. M. Post V. Schusdziarra P. N. Maton R. F. Arakaki

With 31 Figures and 6 Tables

Springer-Verlag
Berlin Heidelberg New York
London Paris Tokyo
Hong Kong Barcelona
Budapest

Dr. Claude Weil
Sandoz Pharma AG
4002 Basel, Switzerland

Professor Dr. Eugenio E. Müller
Dipartimento di Farmacologia
Facoltà di Medicina e Chirurgia
Università degli Studi di Milano
Via Vanvitelli, 32
20129 Milan, Italy

Professor Dr. M. O. Thorner
Dept. of Internal Medicine
School of Medicine
University of Virginia
Charlottesville
Virginia 22908, USA

Cover picture: Three-dimensional view of the brain with special emphasis on the basal ganglia. Illustration by Jack Haley, produced on a Sandoz Scholarship in the Department of Art as Applied to Medicine of the University of Toronto (Chairman: Linda Wilson-Pauwels).

Volume 1: The Dopaminergic System
© Springer-Verlag Berlin Heidelberg 1985

Volume 2: Transmitter Molecules in the Brain
© Springer-Verlag Berlin Heidelberg 1987

Volume 3: The Role of Brain Dopamine
© Springer-Verlag Berlin Heidelberg 1989

ISBN 3-540-54569-7 Springer-Verlag Berlin Heidelberg New York
ISBN 0-387-54569-7 Springer-Verlag New York Berlin Heidelberg

Library of Congress Cataloging-in-Publication Data
Somatostatin/with contributions by Y.C. Patel ... [et al.]
(Basic and clinical aspects of neuroscience : vol.4)
Includes bibliographical references and index.
ISBN 3-540-54569-7 (alk. paper)
ISBN 0-387-54569-7 (alk. paper)
1. Somatostatin. I. Patel, Yogesh C. II. Series. [DNLM: 1. Somatostatin. W1 BA813S v. 4] QP572.S59S66 1992 612.4'05–dc20
DNLM/DLC

Typesetting, printing, and binding: Appl, Wemding
21/3145-543210 – Printed on acid-free paper

Preface

This fourth volume of *Basic and Clinical Aspects of Neuroscience* is devoted to somatostatin (also called somatotropin release-inhibiting hormone, SRIH, abbreviated SS in this volume), a peptide first isolated from hypothalamic tissue and detected soon afterwards in the gastrointestinal tract and pancreatic islets.

Other regulatory peptides have also been found in both the central nervous system and the gastroenteropancreatic system, e.g. substance P, neurotensin, bombesin, enkephalin, and vasoactive intestinal peptide. They are thus known collectively as brain-gut peptides.

As SS is a member of this class, it is considered in this issue with regard to both of these systems. Moreover, SS is dealt with from various angles: its physiological roles, its participation in pathological processes, and the therapeutic uses of an SS analogue (the main reason for using an analogue rather than SS is the extremely short half-life of the natural compound).

Professor E. Flückiger, the first Managing Editor of this series, has chosen to retire. We sincerely thank him for his unstinting efforts in the planning and successful completion of Volumes 1–3. Professor E. E. Müller (Milan) and Professor M. O. Thorner (Charlottesville) have fortunately remained on the editorial board and smoothed my path, for which I am greatly indebted to them.

We hope that this volume of *Basic and Clinical Aspects of Neuroscience,* with its broad scientific and clinical scope, will be received with the same interest and enthusiasm as the first three volumes.

Basel, May 1992

C. Weil
Managing Editor

Contents

General Aspects of the Biology and Function of Somatostatin
Y. C. Patel

Molecular Heterogeneity of Somatostatin 1
Evolution of Somatostatin Genes and Gene Products 3
Anatomical Distribution of Somatostatin Cells 4
Somatostatin in the Blood and Other Body Fluids 5
Actions of Somatostatin 6
Molecular Model of Somatostatin Action 8
  Somatostatin Receptor Subtypes 8
Regulation of Somatostatin Secretion and Gene Expression 9
  Regulation of Secretion 9
  Regulation of Gene Expression 9
Physiological Significance of Somatostatin 11
Somatostatin in Disease 13
References 14

Somatostatin Receptors in the Central Nervous System
J. Epelbaum

Introduction 17
Somatostatin Receptors 17
  In Vitro Binding Assays 17
  Structural Characterization 17
  Localization of Somatostatin Receptors by Radioautography 20
  Mechanisms of Action 21
  Regulation by Hormones and Somatostatin 23
  Multiple Effects 23
Somatostatin Receptors in Pathology 23
Conclusion 25
References 25

Somatostatin in Neuropsychiatric Disorders
D. R. Rubinow, C. L. Davis, and R. M. Post

Introduction 29
Neuropsychiatric Disease-Related Alterations 29
Alzheimer's Disease 29
Depression 30
Central Nervous System Effects of Somatostatin 33
Central Nervous System Localization 33
Neurophysiological Actions 35
Interaction with Neuroregulators 35
Behavioral Actions 35
Somatostatin in Affective Disorder: Clinical Relevance 36
Endocrine Correlates 36
Role in Seizure Disorders and Effects of Psychopharmacological Agents . . . 37
Conclusions 38
References 38

The Physiological Role of Somatostatin in the Regulation of Nutrient Homeostasis
V. Schusdziarra

Introduction 43
Release of Somatostatin 43
Basal Somatostatin Release 43
Postprandial Somatostatin Release 44
Mechanisms of Somatostatin Release 44
Cephalic Phase 44
Gastric Phase 45
Neural Mechanisms During the Gastric Phase 46
Intestinal Phase 46
Intestinal Phase and Gastric Somatostatin 46
Intestinal Phase and Pancreatic Somatostatin 47
Physiological Effects of Somatostatin 47
Somatostatin and Gastric Functions 47
Somatostatin and Pancreatic Functions 49

Exocrine Pancreas 49
Endocrine Pancreas: Effects on the Exocrine Pancreas 49
Somatostatin and Nutrient Entry into the Circulation 50
References 51

Therapeutic Use of Somatostatin and Octreotide Acetate in Neuroendocrine Tumors
P. N. MATON and R. F. ARAKAKI

Pharmacology of Octreotide Acetate. 55
The Effects of Somatostatin and Octreotide on the Pituitary 56
Effects on Pituitary Tumors 56
GH-Secreting Tumors 56
TSH-Secreting Tumors 58
Other Pituitary Tumors 58
Side Effects of Octreotide 58
The Effects of Somatostatin and Octreotide on Neuroendocrine Tumors of the Gut 59
The Carcinoid Syndrome 60
Insulinomas 60
Gastrinomas 60
VIPomas 60
Glucagonomas 61
GHRHomas 61
Cushing's Syndrome 62
Nonfunctioning and Other Rare Neuroendocrine Tumors. 62
Effects of Octreotide on Tumor Size 63
Unresolved Issues 63
References 63

General Aspects of the Biology and Function of Somatostatin

Y. C. Patel

Fraser Laboratories, Departments of Medicine, Neurology, and Neurosurgery, McGill University, Royal Victoria Hospital and Montreal Neurological Institute, Montreal, Quebec, Canada H3A 1A1

Biological activity now recognizable as that of somatostatin (SS) was first encountered in 1968 by Krulich et al. [33] during attempts to screen hypothalamic extracts for growth hormone (GH) releasing activity. The workers identified a GH-inhibitory substance, characterized it as a low-molecular-weight basic peptide, and were able to localize the biological activity to the median eminence and anterior hypothalamic area. A year later, Hellman and Lernmark [23] reported the presence of a potent insulin-inhibitory factor in extracts of pigeon pancreatic islets. These two apparently unrelated observations were to come into sharp focus in 1973 with the discovery by Brazeau et al. [10] of the tetradecapeptide SS-14 as the hypothalamic GH-inhibitory substance, a signal achievement duly recognized by the award of the Nobel Prize to Guillemin in 1977. Subsequent studies revealed that SS is not produced only in the hypothalamus but also occurs throughout the central nervous system (CNS), in peripheral neurons, in the gastrointestinal (GI) tract and in the pancreatic islets of Langerhans. Furthermore, SS-like immunoreactivity is heterogeneous and is distributed in many tissues in both vertebrate and invertebrate species of the animal kingdom and in the plant kingdom. Extensive investigation of SS has shown that its wide anatomical distribution is paralleled by an equally broad spectrum of biological effects.

Today, SS is best regarded as a phylogenetically ancient, multigene family of peptides with two important bioactive products, namely SS-14, the form originally identified in the hypothalamus, and SS-28, a congener of SS-14 extended at the N terminus that was discovered subsequently. These two hormones are produced in various proportions by different SS cells and act either locally on neighbouring cells or more widely through the circulation to regulate such diverse physiological processes as glandular secretion, neurotransmission, smooth muscle contractility, nutrient absorption and cell division. The physiological role of hypothalamic SS in the regulation of GH and thyroid-stimulating hormone (TSH) secretion is well established. The SS peptides also appear to be physiological regulators of islet cells and many GI functions, and may be of considerable importance in the pathophysiology of diseases such as diabetes mellitus, Alzheimer's disease, Huntington's disease and epilepsy. This chapter will present an overview of the basic biology of SS; for a more comprehensive coverage of different aspects of the topic, the reader is referred to several earlier reviews and monographs [16, 47, 56–58, 65].

Molecular Heterogeneity of Somatostatin

Like other protein hormones, SS is synthesized as part of a large precursor molecule (prepro-SS) that is rapidly cleaved into the prohormone (pro-SS) form and processed enzymatically to yield several mature products. Mammalian pro-SS, a 10-kilodalton (kDa) molecule of 92 amino acids, is processed both at the C-terminal segment and at the N-terminal segment of the molecule to yield the two bioactive forms SS-14 and SS-28, together with SS-$28_{[1-12]}$, pro-$SS_{[1-10]}$ (antrin), pro-$SS_{[1-76]}$ (8 kDa) and pro-$SS_{[1-63]}$ (7 kDa), as shown in Fig. 1 [5, 20, 50, 63]. These mature products are generated through proteolytic cleavage at three characteristic sites: a dibasic (Arg-Lys) and a monobasic (Arg) site at the C-terminal segment that generate SS-14 and SS-28, respectively, and a monobasic (Lys) site at the N-terminal region that produces pro-$SS_{[1-10]}$ (Fig. 1). The cleavage sites are highly conserved during vertebrate evolution and are important for regulating the biosynthesis of the products generated.

Mammalian tissues contain various mixtures of the different molecular forms of SS [59]: SS-14 predominates in neural tissues and is virtually the only form in the retina, peripheral nerves, pancreas and stomach; SS-28 accounts for about 20%–30% of total SS-like immunoreactivity in the brain; and within the gut, mucosal SS cells elaborate mainly SS-28 and constitute the largest group of SS-28-producing cells in the body. By contrast, SS in enteric neurons consists of SS-14, in keeping with the preferred neural expression of this molecular form. SS-$28_{[1-12]}$ is found in high concentrations in the pancreas, brain and stomach [50]. Despite its name, antrin is found in large amounts not only in the antrum but also in the rest of the stomach, in the brain and in the intestine [63].

Among the six known cleavage products of SS (Fig. 1), only SS-14 and SS-28 interact with classical SS receptors to evoke SS-like biological effects. The remaining peptides are devoid of any known biological activity, and their function as putative mature products of pro-SS processing remains un-

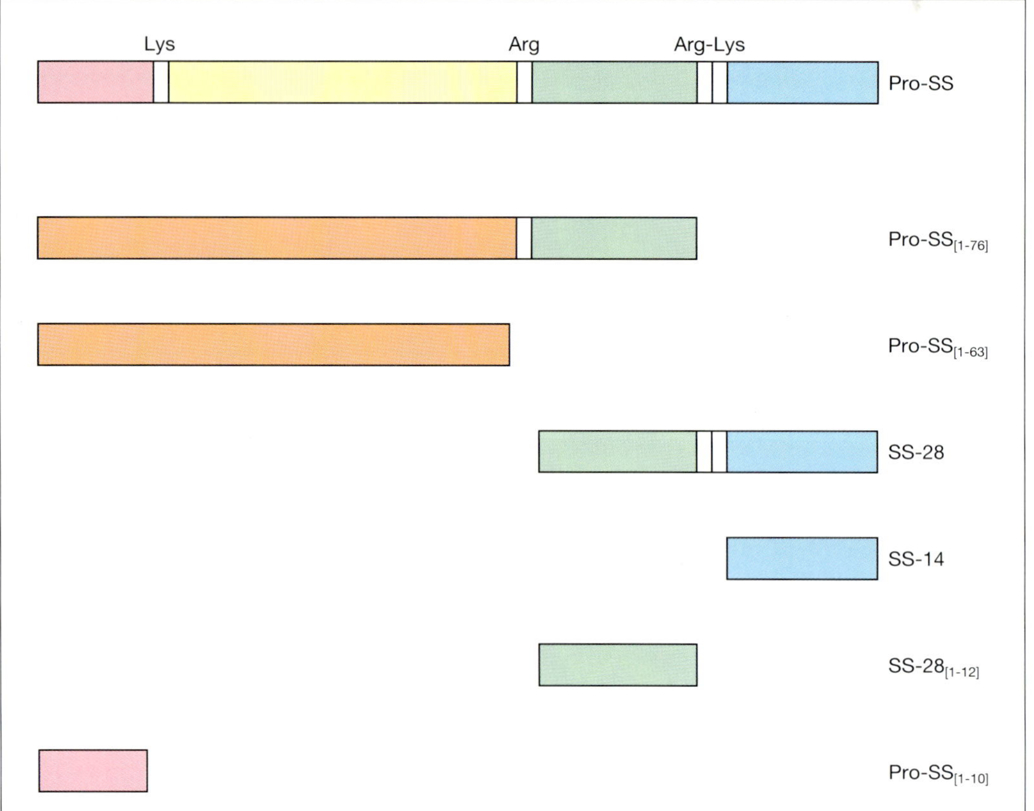

Fig. 1. Schematic representation of mammalian pro-SS and its cleavage products. Processing occurs at three sites marked by the presence of either paired (Arg-Lys) or single (Lys or Arg) basic amino acid residues. Although each of these forms is capable of release from SS cells, only SS-14 and SS-28 are biologically active. (From [63])

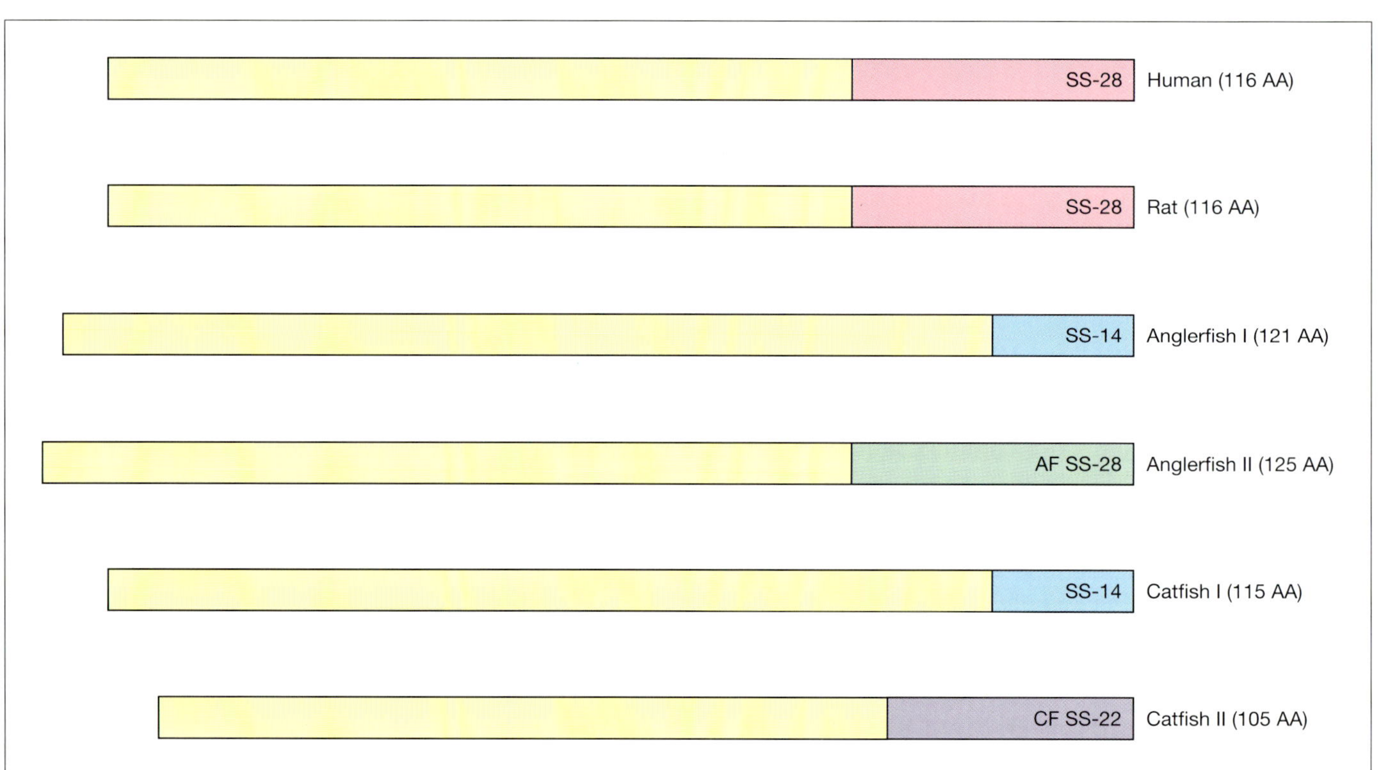

Fig. 2. Schematic depiction of six prepro-SS molecules that have been sequenced so far by cloning techniques (AA amino acid residues). The biologically active SSs in each instance are located at the C-terminal segment of the precursor and consist of SS-14 and SS-28 in rat and humans, SS-14 and anglerfish SS-28 in anglerfish, and SS-14 and catfish SS-22 in catfish. Fish possess two separate SS genes that encode for either SS-14 or a SS-28-type molecule. During vertebrate evolution, one of the two SS genes has become extinct; the single remaining SS gene in mammals expresses both SS-14 and SS-28. See also Fig. 3

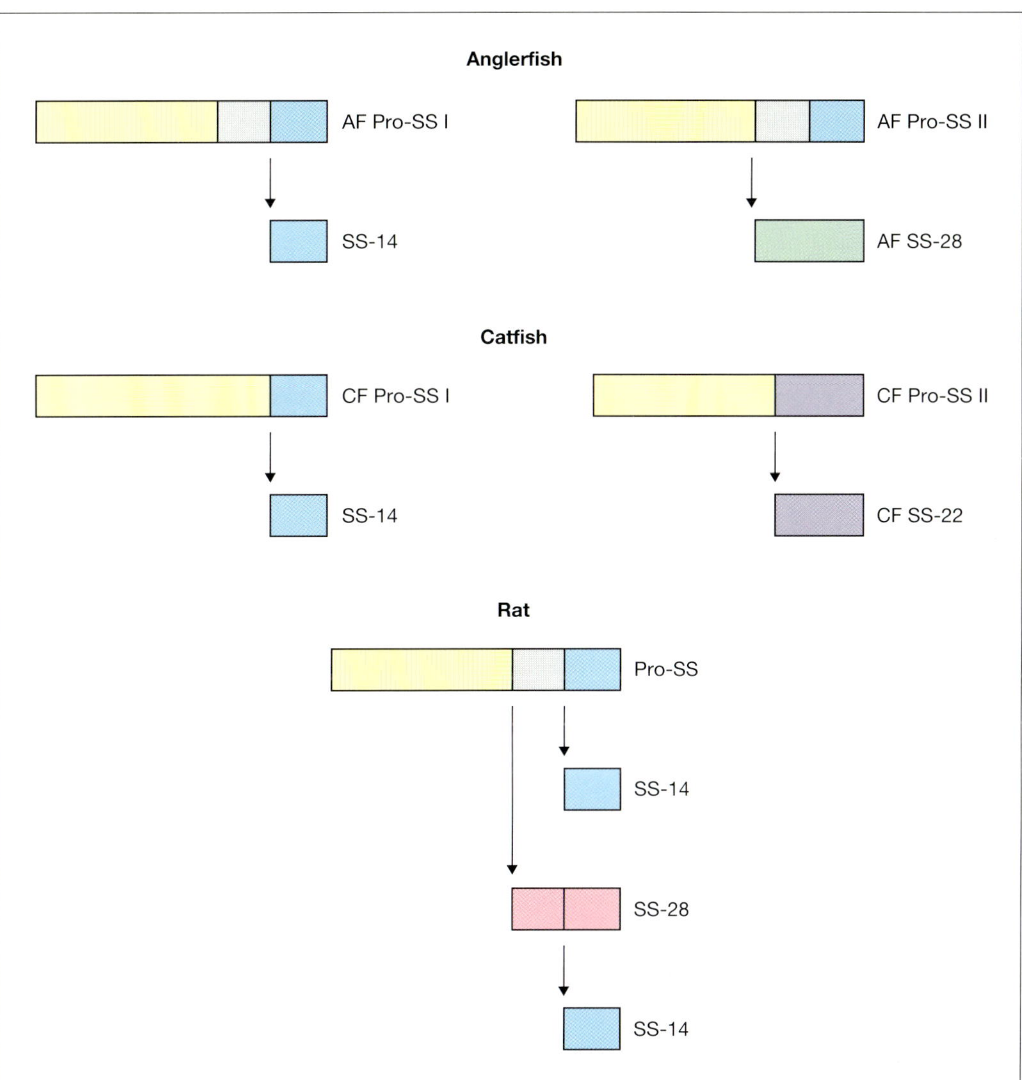

Fig. 3. *Different pathways for the maturation of fish and mammalian pro-SSs. In the anglerfish, two separate precursors for SS are independently processed to either SS-14 or anglerfish SS-28 (a molecule homologous to mammalian SS-28). Likewise, the two catfish pro-SSs yield SS-14 and catfish SS-22 (a peptide distantly related to mammalian SS-28), respectively, as separate products. In contrast, the single mammalian pro-SS (typified by the rat) is cleaved to both SS-14 and SS-28. Although SS-14 is derived principally by direct processing of pro-SS, some SS-14 may originate from further breakdown of SS-28*

certain. However, the total conservation of the amino acid structure of pro-SS$_{[1-10]}$ from fish to mammals suggests an important biological role for this molecule, perhaps as a recognition signal on pro-SS for targeting the prohormone to its intracellular sites of processing. The 8-kDa and 7-kDa forms are possibly just functionless residual portions of the pro-SS molecule after the cleavage of SS-14 and SS-28, comparable for instance to the connecting peptide (C-peptide) moiety in the proinsulin molecule.

Evolution of Somatostatin Genes and Gene Products

Much of the initial work on the structural characterization of SS precursors and on SS biosynthesis was carried out in the fish pancreas. This is because the pancreata of certain teleosts such as the anglerfish and catfish, unlike those of other species, contain a high ratio of endocrine to exocrine tissue and provide a rich source of pure islet cells. To date, six separate prepro-SS molecules have been identified and sequenced [18, 20, 24, 38, 75, 82], as shown in Fig. 2. Of these, the human and rat prepro-SSs are virtually identical, differing in only four amino acid residues. In the fish, however, four separate prepro-SSs have been identified; these four precursors are the products of two separate pairs of SS genes, a pair each for the two fish species. By contrast, all evidence so far points to a single SS gene in mammals [42]. The fish precursors exhibit about 40 % amino-acid sequence homology with their mammalian counterparts [2]. The biologically active hormone is in each case located at the C terminus of the prepropeptide and consists of SS-14 and SS-28 in mammals and SS-14, anglerfish SS-28 and catfish SS-22 in fish. The SS-14 sequence is totally conserved from fish to mammals, whereas mammalian SS-28 shares only 40 %–66 % homology with its two fish counterparts.

There are major differences in the way in which mammalian and fish pro-SSs are processed enzymatically (Fig. 3). Whereas the single mammalian pro-SS molecule is cleaved to generate the two mature forms SS-14 and SS-28, SS-14 and anglerfish SS-28 are the two separate and independent products of anglerfish pro-SS I and anglerfish pro-SS II, respectively [37, 44]. Likewise, the catfish precursors yield SS-14 and catfish SS-22, respectively, as their mature products. Recent immunocytochemical and in situ hybridization data indicate that the two catfish and the two anglerfish gene products are located in two distinct populations of islet D cells and suggest that each of the two fish genes is expressed in separate cells [41, 74]. Nucleotide sequence homologies between the fish and human SS genes suggest that, of the two fish SS genes, the one that codes for SS-28/SS-22 has become extinct at some point in vertebrate evolution and that the SS-14 gene of the fish has evolved into the mammalian form and assumed the function of both genes [2]. Given the contrasting paradigms between fish – with two SS genes expressed in two separate populations of SS cells, two precursors and two products – and mammals – with a single gene, a single precursor and two products that appear to be located in the same cells – a great deal still needs to be known to bridge the gap in our understanding of the evolution, cellular expression and function of the SS genes from fish to mammals and to explain the apparent loss of the second SS gene.

Anatomical Distribution of Somatostatin Cells

SS-producing cells occur at high densities throughout the central and peripheral nervous systems, in the endocrine pancreas and in the gut and in small numbers in the thyroid, adrenals, submandibular glands, kidneys, prostate and placenta (Table 1) [47, 64]. SS-containing nerve fibres have been detected in the heart. The typical morphological appearance of a SS cell is that of a neuron with multiple branching processes or of a secretory cell often having short cytoplasmic extensions (D cells). Within the hypothalamus, the most prominent collection of SS-positive nerve-cell bodies lies in the anterior periventricular region [17, 27, 32]. The periventricular SS cells are located close to the third ventricle, in three or four layers that are parallel to the ventricular wall within an ovoid area that extends from the preoptic nucleus to the rostral margin of the ventromedial nucleus [17, 32]. Axons from these cells sweep laterally from the periventricular region and run caudally through the hypothalamus to form a discrete pathway towards the midline that enters the median eminence at the level of the ventromedial nucleus. Within the median eminence, somatostatinergic nerve endings extend in a compact band throughout the zona externa. A proportion of the fibres from this pathway course through the neural stalk and terminate in the neurohypophysis. The anterior hypothalamic periventricular SS pathway to the median eminence accounts for about 80% of SS immunoreactivity in the hypothalamus [13]. SS perikarya in low to moderate densities occur in several other hypothalamic regions, notably the paraventricular, arcuate and ventromedial nuclei [17, 27, 32]; these cells, however, do not appear to make significant contributions to median eminence SS. In addition to those in the median eminence, axons from the periventricular neurons project widely within the hypothalamus and via long extensions to extrahypothalamic structures, notably in the limbic system [32].

Table 1. *Localization of somatostatin*

Body region	*Type of cells*	*Location*
Major sites		
Nervous system	Neurons	Hypothalamus
		Cerebral cortex
		Limbic system
		Basal ganglia
		Major sensory systems
		Spinal cord
		Dorsal root ganglia
		Autonomic ganglia
Pancreas	D cells	Islets
Gut	D cells	Mucosal glands
	Neurons	Submucous and myenteric plexuses
Minor sites		
Adrenal	–	Scattered medullary cells
Placenta	–	Cytotrophoblasts in chorionic villi
Reproductive organs	–	Testis, epididymis, prostate
Submandibular gland	D cells	Scattered ductal cells
Thyroid	C cells	Scattered parafollicular cells (coexisting with calcitonin)
Urinary system	–	Scattered cells in renal glomerulus and collecting ducts

From [47]

Outside the hypothalamus, SS-positive neurons and fibres are abundantly dotted throughout the CNS, with the notable exception of the cerebellum [17, 27]. Brain regions rich in SS cells include the deeper layers of the cortex, all limbic structures, the striatum, the periaqueductal central gray and all levels of the major sensory systems. For instance, somatostatinergic neurons occur at initial sites of synaptic processing in somatic sensory systems and at several levels in special sensory systems such as the olfactory, visual and auditory systems. The approximate relative amounts of SS in the major regions of the brain are as follows: cerebral cortex 49%, spinal cord 30%, brain stem 12%, hypothalamus 7%, olfactory lobe 1% and cerebellum 1% [55]. Like other neuropeptides, SS coexists with a number of peptides and transmitters [16]. The earliest example of such coexistence was that with noradrenaline in postganglionic neurons of the inferior mesenteric ganglion; more recent examples are those with γ-aminobutyric acid (GABA) in a subpopulation of neocortical neurons, with neuropeptide Y (NPY) in cortical and striatal neurons, and with enkephalin in the median eminence of some species.

SS cells in the pancreas are confined to the islets of Langerhans, where they occur as D cells [3]. In the fetus and neonate, SS cells are the second most abundant islet-cell type after insulin cells, accounting for up to 40% of the total endocrine cell population; in the adult, however, they make up only about 3% of islet cells. Islet D cells are characteristically adjacent to glucagon cells and pancreatic polypeptide cells and are located in the peripheral mantle zone, although there are species-specific variations in this pattern of distribution. GI SS cells are of two types: D cells and neurons intrinsic to the gut [34]. The former are located in mucosal glands from the cardiac portion of the stomach to the rectum, their highest concentration being in the antrum; the latter populate both the submucous and myenteric plexuses in all segments of the GI tract [34]. In the thyroid, SS coexists with calcitonin in a subpopulation of C cells [64]. Scattered SS cells are found in many other organs including the adrenal medulla, testes, prostate, submandibular glands, kidneys and placenta [64, 71]. In the rat, the gut accounts for about 65 % of total body SS, the brain for 25 %, the pancreas for 5 % and the remaining organs for 5 % [59].

Somatostatin in the Blood and Other Body Fluids

Both SS-14 and SS-28 are detectable in the blood [59, 76], the main source of circulating SS being the GI tract. Circulating

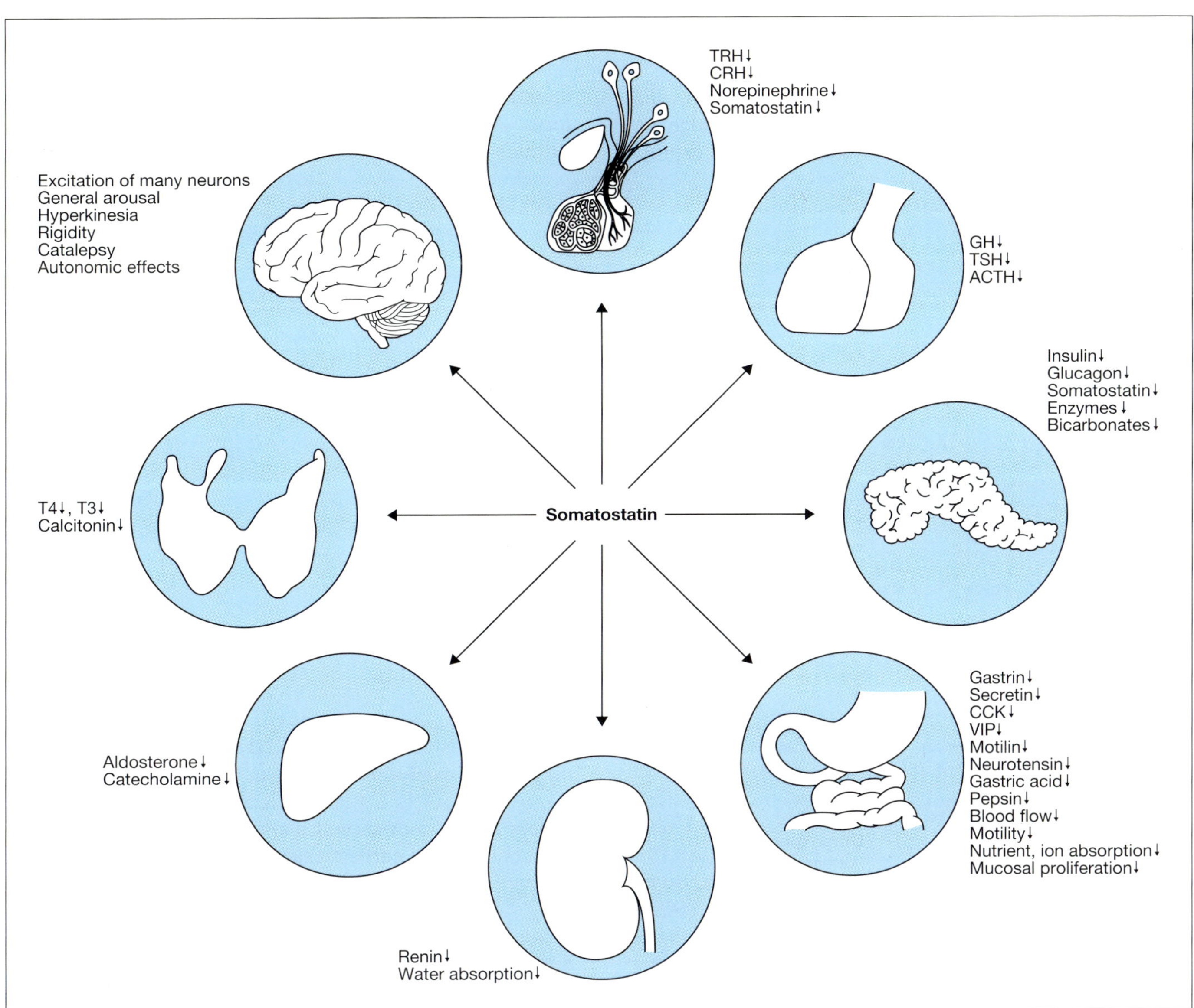

Fig. 4. *Principal actions of SS. It inhibits both the basal and the stimulated secretion of GH, TSH, and islet hormones. It has no effect on LH, FSH, prolactin or ACTH in normal subjects. It does, however, suppress elevated ACTH levels in Addison's disease and in ACTH-producing tumors. In addition, it inhibits the basal and the TRH-stimulated release of prolactin in vitro and diminishes elevated prolactin levels in acromegaly. In the GI tract, SS inhibits the release of virtually every gut hormone that has been tested. It has a generalized inhibitory effect on gut exocrine secretion (gastric acid, pepsin, bile, colonic fluid) and suppresses motor activity generally as well through inhibition of gastric emptying, gallbladder contraction and small intestine segmentation. SS, however, stimulates MMC (migrating motor complex) activity. The effects of SS on the thyroid include the inhibition of the TSH-stimulated release of T_4 and T_3. The adrenal effects consist of the inhibition of angiotensin-II-stimulated aldosterone secretion and the inhibition of acetylcholine-stimulated medullary catecholamine secretion. In the kidneys, SS inhibits the release of renin stimulated by hypovolaemia and inhibits ADH-mediated water absorption.* ***ACTH****, adrenocorticotropic hormone;* ***ADH****, antidiuretic hormone;* ***CCK****, cholecystokinin;* ***CRH****, corticotropin-releasing hormone;* ***GH****, growth hormone;* ***T_3****, triiodothyronine;* ***T_4****, thyroxine;* ***TRH****, thyrotropin-releasing hormone;* ***TSH****, thyroid-stimulating hormone;* ***VIP****, vasoactive intestinal peptide*

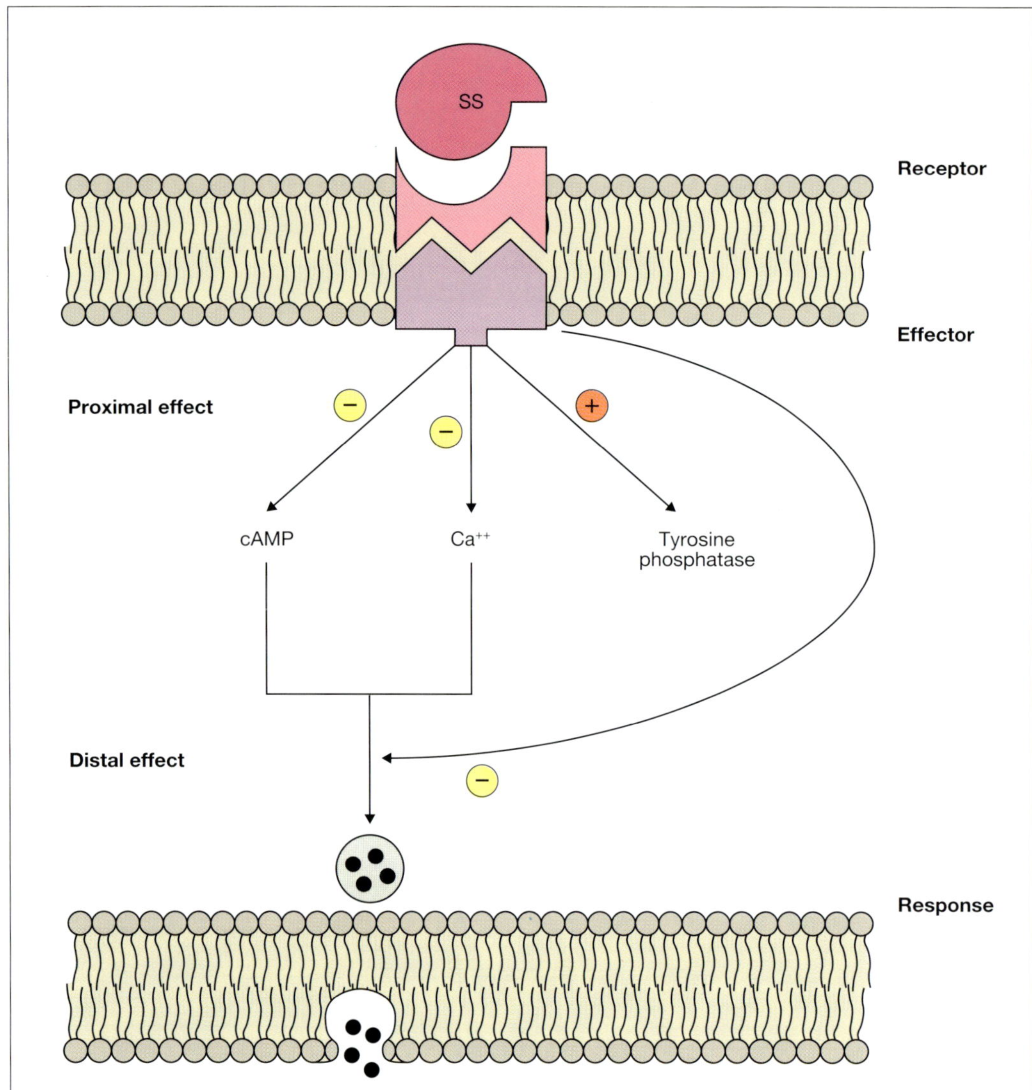

Fig. 5. Schematic model of SS receptor function. The receptor is linked to several effector systems that generate different transmembrane signals. Activation of the receptor by SS induces (1) inhibition of membrane adenylyl cyclase and a fall in cAMP levels, (2) activation of membrane ion channels (K^+ and Ca^{2+}) leading to a drop in intracellular free Ca^{2+} concentration, (3) stimulation of a membrane-associated tyrosine phosphatase, and (4) inhibition of exocytosis by a so-called "distal effect" independent of any changes in the levels of intracellular second messengers such as cAMP or Ca^{2+}. (From [49])

SS is rapidly inactivated by the liver and kidneys. The plasma half-life of SS-14 in man is 2–3 min, whereas SS-28 is slightly more resistant to inactivation. Fasting plasma concentrations of SS range from 5 to 18 pmol/l, and these levels double in response to the ingestion of a mixed meal. The bioactive circulating forms consist of SS-14, des-Ala[1]-SS-14 (a postsecretory conversion product of SS-14) and SS-28 [76]. With few exceptions, fluctuations in SS levels in the peripheral plasma are very small. The main clinical utility of plasma measurements lies in the diagnosis of SS-producing tumours, which are usually associated with marked hypersomatostatinaemia.

SS secreted into the cerebrospinal fluid (CSF) probably emanates from all parts of the brain [54]; owing to its greater stability in the CSF, its concentration is twice that in the general circulation. Significant amounts are excreted in the urine (4–6 pmol/l), where it is present as intact SS-14, intact SS-28 and various metabolites [62]. Semen contains very high concentrations of SS-like immunoreactivity [71]: the spermatic levels in humans are 200-fold those in plasma and consist almost only of SS-28. Amniotic fluid is rich in SS, originating mainly from the fetus.

Actions of Somatostatin

Since its discovery 20 years ago, no other feature of SS has captured the minds of both basic and clinical scientists more than its extraordinary range of biological effects [49, 64]. This interest has been enhanced with the recent advent of clinically useful SS analogues and the resulting need for a detailed understanding of the pharmacology of SS.

Along with its wide anatomical distribution, SS acts on multiple targets including the brain, gut, pituitary, endocrine and exocrine pancreas, adrenals, thyroid and kidneys. As shown in Fig. 4, its actions include inhibition of virtually every known endocrine and exocrine secretion and various neurotransmitters, behavioural and autonomic effects of

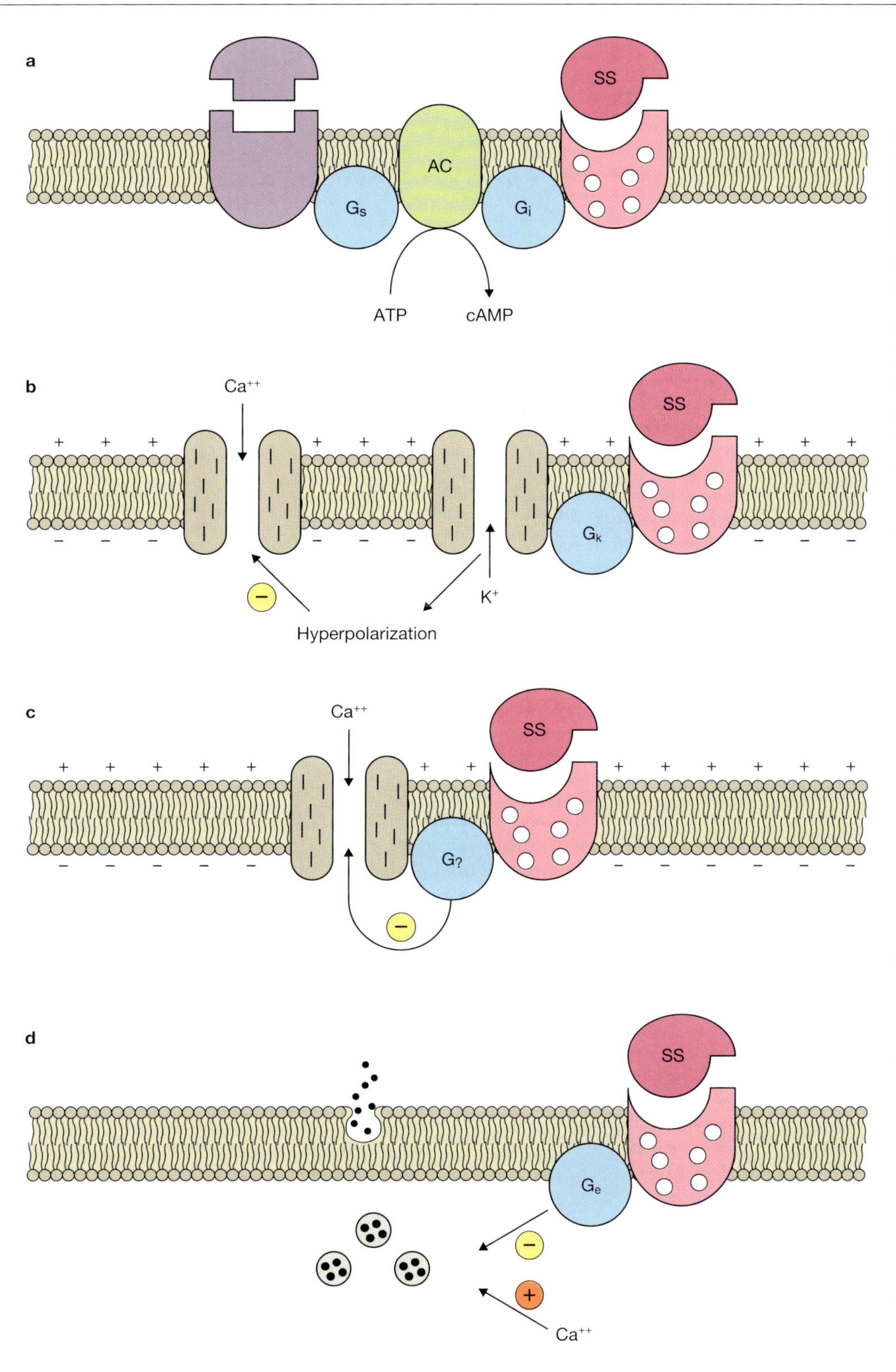

***Fig. 6 a–d.** Schematic illustration of the function of four transmembrane signalling pathways linked to the SS receptor that act on the secretory process. **a** Receptor coupling to adenylyl cyclase **(AC)** via the inhibitory guanine nucleotide **(GTP)** binding protein G_i. The activity of AC is regulated by stimulatory and inhibitory receptors coupled to the enzyme by the GTP-binding proteins G_s and G_i, respectively. The activation of the SS receptor is associated with the inhibition of stimulated AC activity and a fall in cAMP levels. **b** Receptor coupling directly to K^+ channels via a GTP-binding protein G_k. Receptor activation leads to an opening of K^+ channels, efflux of K^+ ions, membrane hyperpolarization (i. e. the cell becomes more negative inside and thus more refractory to depolarization) and a secondary reduction in intracellular Ca^{2+} due to inhibition of the normal depolarization-induced influx of Ca^{2+} ions via voltage-sensitive Ca^{2+} channels. **c** Receptor coupling directly to Ca^{2+} channels via a suspected GTP-binding protein **G?** Receptor activation blocks Ca^{2+} influx from voltage-dependent Ca^{2+} channels and causes a fall in intracellular Ca^{2+} concentrations. **d** Receptor coupling to exocytotic granule via a putative GTP-binding protein G_e. The underlying molecular mechanisms here are poorly understood. (From [49])*

centrally administered SS, and effects on GI and biliary motility, vascular smooth muscle tone and intestinal absorption of nutrients and ions [47, 64]. These actions are mediated through high-affinity membrane receptors that occur at various densities in all target tissues [49, 66, 78]. At a cellular level, the broad array of biological actions can be resolved into four processes that are regulated by the peptide, namely neurotransmission, glandular secretion, smooth muscle contractility and cell proliferation. The effects of SS on smooth muscle contraction can be largely explained through modulation of acetylcholine release, a secretory process, so an understanding of the molecular basis of SS action requires an explanation of how it alters cell excitability, the secretory process and cell growth.

Molecular Model of Somatostatin Action

Critical to the action of SS is the SS receptor, which, like other hormone receptors, subserves two functions: (1) to recognize the ligand and bind it with high affinity and specificity, and (2) to generate a transmembrane signal that evokes a biological response (Fig. 5). Five main receptor-linked membrane signalling pathways have been identified (Figs. 5, 6):

1. Receptor coupling to adenylyl cyclase
2. Receptor coupling to K^+ channels
3. Receptor coupling to Ca^{2+} channels
4. Receptor coupling to exocytotic vesicles
5. Receptor coupling to tyrosine phosphatase

The first four pathways are responsible for SS-receptor-mediated effects on the secretory process and each require the interaction of the receptor with a guanine nucleotide (GTP) binding protein (G-protein). Receptor activation is associated with prompt reductions in two key intracellular mediators, cyclic adenosine monophosphate (cAMP) and Ca^{2+}, due to receptor-linked effects on the plasma membrane enzyme adenylyl cyclase and on K^+ and Ca^{2+} ion channels [26, 49, 73, 89] (Figs. 6 A–C). The changes in cAMP and Ca^{2+} occur independently of each other. Available evidence does not suggest an effect of SS on diacylglycerol or inositol phosphate formation. These so-called proximal effects help to explain part of the inhibitory action of SS on hormone secretion (Fig. 5). SS, however, like a number of other inhibitory receptors (e.g. α_2-adrenergic, M_2 muscarinic), is also known to act at a step distal to cAMP generation and Ca^{2+} entry [36, 87]. Thus, SS blocks hormone secretion stimulated directly by various second messengers such as cAMP, inositol triphosphate or diacylglycerol or by elevating intracellular Ca^{2+} concentrations [36, 87]. These observations clearly indicate that – independently of any effect on cAMP, Ca^{2+} or indeed any known second messenger – SS is able to inhibit hormone secretion via a distal effect targeted through the receptor on the exocytotic process (Fig. 6 D). The fifth effector system is a tyrosine phosphatase recently shown to be stimulated by SS receptor activation, with resulting dephosphorylation and inactivation of tyrosine kinases such as the epidermal growth factor (EGF) receptor [35] (Fig. 5). Since tyrosine kinases play an important role in cell division, their inactivation by SS may explain the antiproliferative actions of the peptide (see also the chapter by Maton and Arakaki, this volume).

From this scheme, it is evident that the SS receptor is linked to multiple signalling pathways via GTP-binding proteins (except for coupling to tyrosine phosphatase, where a G-protein mediation is currently unknown). What remains to be determined now is whether these different signalling pathways are all expressed in a given target cell or whether they are cell-specific, and what their relative function is with respect to the overall biological response.

Somatostatin Receptor Subtypes

Pharmacological studies have shown that the SS receptor is heterogeneous and may exhibit subtypes specific for SS-14 or SS-28 [49, 78]. For instance, SS-14 and SS-28 exert differential effects on their target cells. SS-28 tends to be relatively more selective than SS-14 in inhibiting the secretion of hypothalamic corticotropin releasing factor (CRF) and vasopressin, pituitary GH and TSH, insulin and products of the exocrine pancreas, whereas SS-14 appears to exert a more potent effect than SS-28 on cerebrocortical neurons,

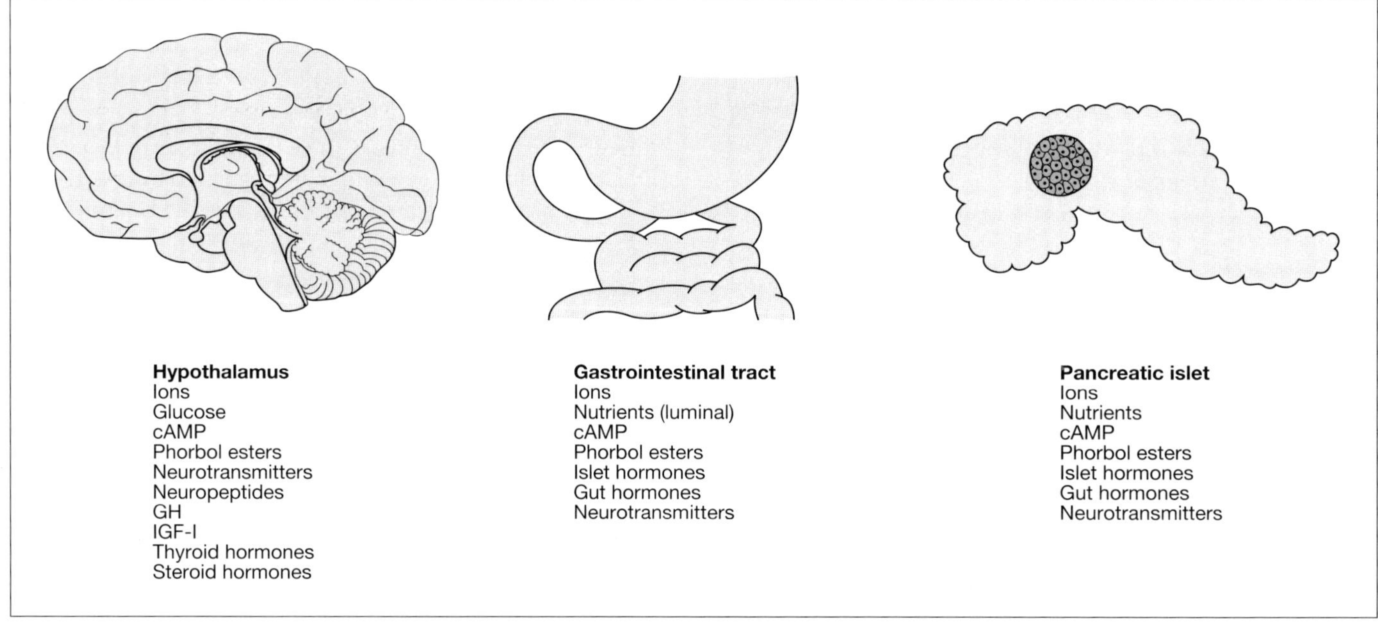

Fig. 7. *Agents that regulate SS secretion and/or gene expression in the hypothalamus, pancreatic islets and gastrointestinal tract*

glucagon and gastric exocrine secretion, GI motility and mesenteric blood flow.

Two populations of SS receptors can be distinguished by their ability or inability to be desensitized upon exposure to SS (homologous desensitization) [49]. Receptors that bind SS-28 preferentially, e. g. in normal pituitary and islet B cells, are downregulated by treatment with SS-14 or SS-28, whereas SS-14-selective receptors, e. g. on islet A cells, are resistant to such desensitization [49]. Receptor subtypes could be explained by multiple receptor proteins, as is typical in other recently characterized receptor systems, or be due to a single receptor protein with different conformations resulting from posttranslational modifications, e. g. glycosylation. However, the recent cloning of several receptor systems typified by receptor subtypes, e. g. the adrenergic, muscarinic, dopamine and tachykinin receptor families, has shown the existence of distinct receptor proteins for each receptor subclass. Furthermore, subtypes in the adrenergic and muscarinic receptor families are selectively coupled via distinct G-proteins to their effector systems.

The biological activity of SS-14 and SS-28 in general correlates with their potency for receptor binding in most systems, a fact suggesting that their selective actions may result mainly from differential interactions with receptors. The recent biochemical characterization of the SS receptor by photoaffinity labelling and solubilization has revealed the presence of several different protein forms, some of which exhibit selectivity for binding to SS-14 or SS-28 [49]. In addition to this heterogeneity at the level of the receptor molecule, there is evidence that in some cellular systems SS-14 and SS-28 can also exert preferential effects on receptor-linked transmembrane signalling systems. For instance, SS-14 and SS-28 exercise opposite effects on K^+ channels in cerebrocortical neurons [84].

All these observations clearly point to multiple receptor proteins and to their differential interaction with transmembrane signalling systems as the basis for SS-receptor subtypes. For further information on SS receptors, see also the chapter by Epelbaum (this volume).

Regulation of Somatostatin Secretion and Gene Expression

Regulation of Secretion

Because SS cells are so widely distributed and interact with many different body systems, it is not surprising to find that the secretion of SS can be influenced by a broad array of secretagogues, ranging from ions and nutrients to neuropeptides, neurotransmitters, classical hormones and growth factors (Fig. 7) [16, 56, 60, 64, 85, 88]. Some of these agents exert common effects on SS cells in different locations, presumably by direct action; others tend to be tissue selective, a fact that can be explained by tissue-selective expression of receptors for the secretagogues on SS cells or by indirect effects through the release of other peptides or transmitters. The release of SS induced by membrane depolarization occurs both from neurons and from peripheral SS cells, e. g. islet or gastric D cells, suggesting that this mode of release is a fundamental property of all SS cells [56, 64]. Nutrients exert tissue-specific effects on SS release, most prominent in islet D cells stimulated by glucose, amino acids and lipids [85]. Hypothalamic SS secretion is inhibited by glucose and is insensitive to aminogenic stimuli, whereas gut SS secretion is triggered by luminal but not circulating nutrients [6, 56, 88].

Virtually every neurotransmitter or neuropeptide tested has been shown to exert some effect on SS secretion, with varying degrees of potency and tissue selectivity. Within this group, glucagon, GH-releasing hormone (GHRH), neurotensin, CRF, calcitonin-gene-related peptide (CGRP) and bombesin are potent stimulators of SS release from several tissue sites, and opiates and GABA generally inhibit SS secretion [16, 56, 60, 64, 85, 88]. Of the various hormones studied, GH and thyroid hormones enhance SS secretion from the hypothalamus; their effect in other tissues has not been adequately investigated [7, 8]. Glucocorticoids exert a dose-dependent biphasic effect on SS secretion, low doses being stimulatory and high doses inhibitory [45]. Insulin stimulates hypothalamic SS release but has an inhibitory effect on the release of islet and GI SS [6, 53, 88]. Finally, growth factors such as insulin-like growth factor I (IGF-I) and cytokines such as interleukin-1 (IL-1) and tumour necrosis factor (TNF) have recently been shown to stimulate SS secretion from brain cells and appear to be forerunners in what will undoubtedly be an expanding list of compounds capable of regulating SS secretion [9, 72].

Regulation of Gene Expression

The SS gene has so far been sequenced in the rat and humans, where it exhibits a simple configuration. The coding region consists of two exons separated by an intron (Fig. 8) [42]. The 5′ upstream region contains three regulatory elements, a TATA and a CAAT box (highly conserved regions of DNA known as *promoters*), and, in between the two, an eight-base-pair palindromic *enhancer* sequence – the cAMP response element (CRE) [21]. The CRE motif was first identified in the SS gene and has since been shown to be present in a large number of genes that are regulated by cAMP [43]. Gene transcription is activated when specific DNA binding proteins interact with the promoter and enhancer elements of DNA. Promoters determine the constitutive or basal level of gene transcription, whereas enhancers confer tissue specificity and mediate the regulated expression of gene transcription, e. g. that induced by extracellular signals such as hormones.

Many of the agents that influence SS secretion are also capable of altering SS gene expression, as assessed by changes in steady-state messenger RNA (mRNA) levels (Fig. 7). For instance, SS secretion and mRNA accumulation are stimulated by GH in the hypothalamus [7, 68] and by IL-1, TNF, and *N*-methyl-D-aspartate (NMDA) receptor agonists in the cerebral cortex [46, 72] and inhibited in the pancreatic islets by insulin [53]. Testosterone and oestradiol

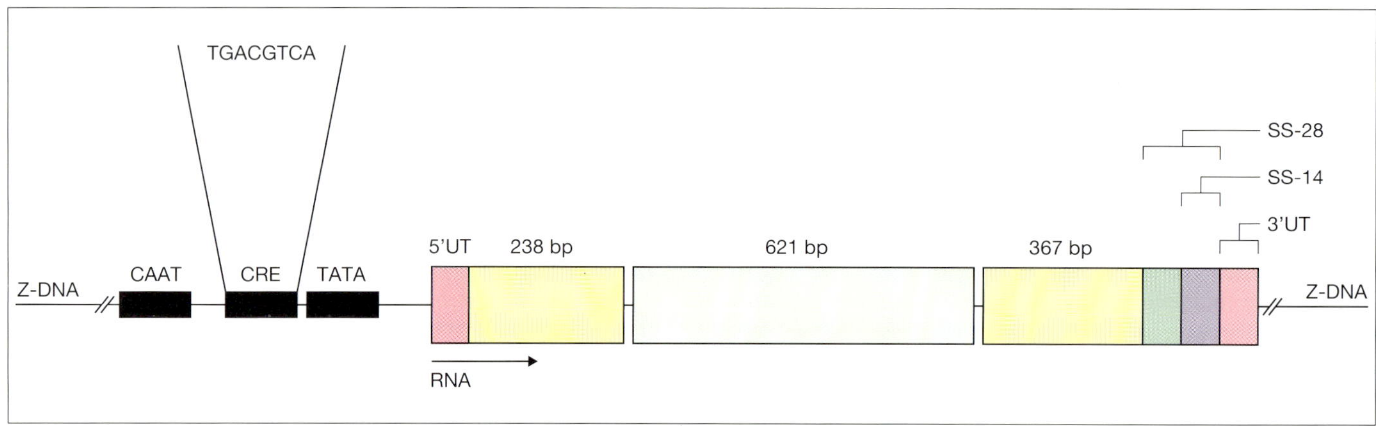

Fig. 8. *Schematic depiction of the rat SS gene. The messenger RNA (mRNA) coding region consists of two exons of 238 and 367 base pairs (bp) separated by an intron of 621 bp. Located upstream (i.e. 5′ end) from the start site of mRNA transcription are three regulatory elements, a TATA and a CAAT box and, between the two, the cAMP response element (CRE). Gene transcription is induced when these regulatory elements interact with specific DNA binding proteins. The gene is flanked at both the 5′ and the 3′ ends by specialized DNA known as zDNA, believed to be additional sites of protein binding and regulation.* ***RNA*** *(arrow), the start site of transcription for mRNA;* ***5′UT, 5′*** *untranslated region;* ***3′UT, 3′*** *untranslated region;* ***SS-14, SS-28,*** *regions in the gene coding for SS-14 and SS-28*

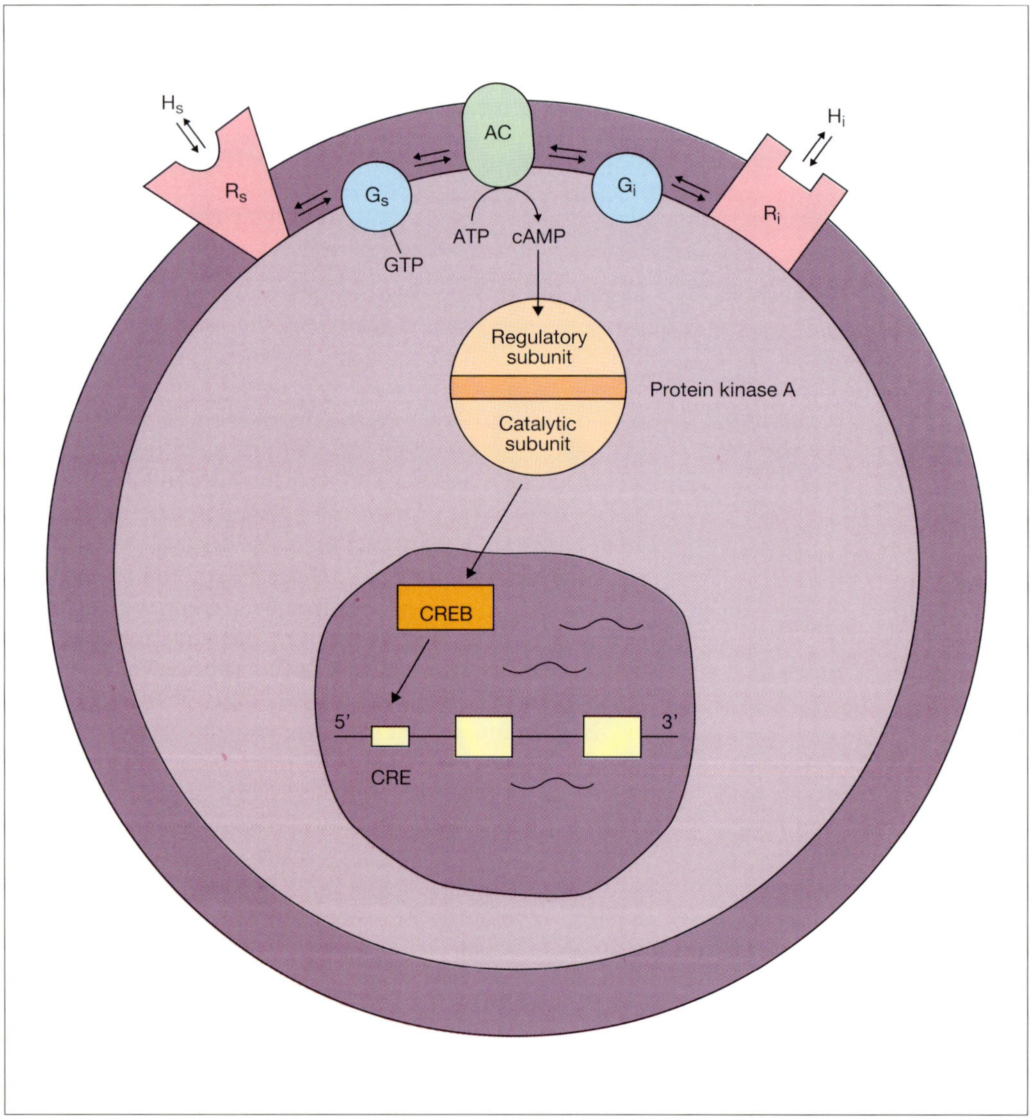

Fig. 9. *Schematic illustration of the steps involved in the transcriptional regulation of the SS gene by the cAMP pathway. Following receptor-induced stimulation of adenylyl cyclase* ***(AC)****, the increase in cAMP levels leads to the release of the active catalytic subunit from the regulatory subunit of the protein kinase A holoenzyme. The catalytic subunit is directly transported to the nucleus, where it phosphorylates the nuclear protein* ***CREB*** *(cAMP response element binding protein). Phosphorylated CREB binds to the cAMP response element* ***(CRE)*** *of the somatostatin gene and stimulates transcription.* ***ATP****, adenosine triphosphate;* ***GTP,*** *guanosine triphosphate;* G_i*, inhibitory GTP-binding protein;* G_s*, stimulatory GTP-binding protein;* H_i*, inhibitory hormone;* H_s*, stimulatory hormone;* R_i*, inhibitory receptor;* R_s*, stimulatory receptor*

both augment hypothalamic SS mRNA [11, 86], and the oestrogen effect is not observed in the cerebral cortex [85].

Among the intracellular mediators known to modulate SS function are ions, cAMP and activators of protein kinase C [56, 60, 85, 88]. Activation of the adenylyl-cyclase-cAMP pathway plays an important role in the stimulation of SS secretion. Such stimulation can be effected physiologically by extracellular agents, e.g. glucagon or GHRH, acting via cell surface receptors coupled to adenylyl cyclase, or pharmacologically through postreceptor and other mechanisms that augment intracellular cAMP levels. cAMP-dependent mechanisms not only stimulate SS secretion but also regulate SS gene transcription (Fig. 9). cAMP activates the transcription of SS and other cellular genes through CRE, which binds a 43-kDa nuclear protein CREB (cAMP response element binding protein) whose transcriptional efficacy is regulated through phosphorylation by the cAMP-dependent enzyme protein kinase A (Fig. 9) [21, 43]. Since cAMP regulates both gene transcription and protein secretion, it could potentially play a role in linking these two cellular processes, its effect on secretion perhaps being mediated through the action of CREB on other cAMP-responsive genes involved in exocytosis. The cAMP system thus emerges as the most prominent intracellular pathway in the SS cell for transmitting external signals directly to the nucleus to activate SS gene transcription. Activators of protein kinase C (a key regulatory enzyme of the inositol phosphate pathway) such as the tumour-promoting phorbol esters can also stimulate transcription of target genes through phosphorylation of nuclear factors comparable to CREB [43]. These agents, however, are without effect on SS gene transcription although they remain potent stimulators of SS secretion [51].

Very little is currently known about the molecular mechanisms underlying the actions of steroid hormones, GH, growth factors, cytokines and NMDA-receptor agonists on SS gene expression. In particular, it will be of considerable interest to determine whether these hormones activate SS gene transcription directly and, if so, whether this action is mediated via the cAMP pathway or another signalling system.

Physiological Significance of Somatostatin

There is growing evidence, both direct and indirect, that SS modulates the physiological function of various target cells. It subserves mainly local regulatory functions, acting as a neurotransmitter or neuromodulator, a neurosecretory substance (i.e. one released directly from nerve axons into the bloodstream, as in the median eminence) or a paracrine/autocrine regulator (local cell-to-cell interaction or self-regulation). Whether circulating SS represents a spill-over of the peptide into the bloodstream after it has already acted at local sites or whether it functions as a true endocrine agent has attracted considerable interest. Because SS cells are widely scattered in the body and because their secretory product is highly labile in the blood and is able to inhibit a wide range of seemingly unrelated endocrine and exocrine cells, it has been argued that SS in the systemic circulation may not have physiological relevance. However, the levels of circulating SS that are achieved physiologically in certain circumstances, e.g. postprandially, are capable of inhibiting the secretion of pituitary or islet hormones and various GI functions [12, 14, 22, 77]. Furthermore, neutralizing the circulating SS by means of antibody augments these functions. In sum, the evidence strongly supports an endocrine role of SS.

There is, at present, good direct evidence of a physiological role for hypothalamic SS in the regulation of GH and TSH secretion by the pituitary [56, 81]. The anterior hypothalamic periventricular somatostatinergic neurons with their projections to the median eminence constitute the final common pathway for the inhibition of GH secretion (Fig. 10). Disruption of this pathway by appropriately placed lesions or by surgical section augments GH and TSH secretion in the rat. Conversely, electrical stimulation of this neuronal system potentiates the release of SS into the hypophyseal portal vessels and inhibits GH release. GH secretion is positively regulated by GHRH produced by arcuate neurons whose axons also project to the median eminence. The SS and GHRH pathways interact with each other not only at their point of convergence at the level of the pituitary but also via direct neural connections within the hypothalamus [25] (Fig. 10). SS thus inhibits GH secretion both by a direct action on the pituitary and indirectly through suppression of GHRH release [29, 80]. The secretion of SS in turn is modulated by GHRH [28] and is subject to positive-feedback regulation by GH (short loop) [7, 56] and IGF-I produced by GH action on the liver (long loop) [9]. A variety of physiological GH responses are orchestrated by SS acting either alone or in concert with GHRH. For instance, normal secretion of GH occurs in a typical pulsatile manner, with high-amplitude secretory bursts at regular intervals during the 24-h period [56, 81, 83]; such episodic secretion is believed to be governed by a characteristic interplay between SS and GHRH, which are released in reciprocal cycles into the hypophyseal portal circulation [61]. Similarly, SS participates in GH regulatory responses to physiological stimuli such as stress, glucose administration or food deprivation.

Because of its extensive extrahypothalamic brain distribution, its effects on the spontaneous electrical activity of neurons, its release from nerve endings in response to depolarization and its behavioural effects, SS has been postulated to serve as a central neurotransmitter or modifier of neuronal function. Given the high concentration of both neuronal elements and SS receptors in limbic, neocortical, striatal and sensory areas, SS appears to be particularly important in modulating functions in these regions [16, 17, 27, 66].

The physiological function of SS in the pancreatic islets of Langerhans has been difficult to characterize [70, 85]. Islet D cells are certainly capable of autoregulating their own secretion and could theoretically influence other islet cells by the paracrine route (i.e. through the interstitium without having passed through the microcirculation) or hormonally via the islet microcirculation (Fig. 11). Indirect evi-

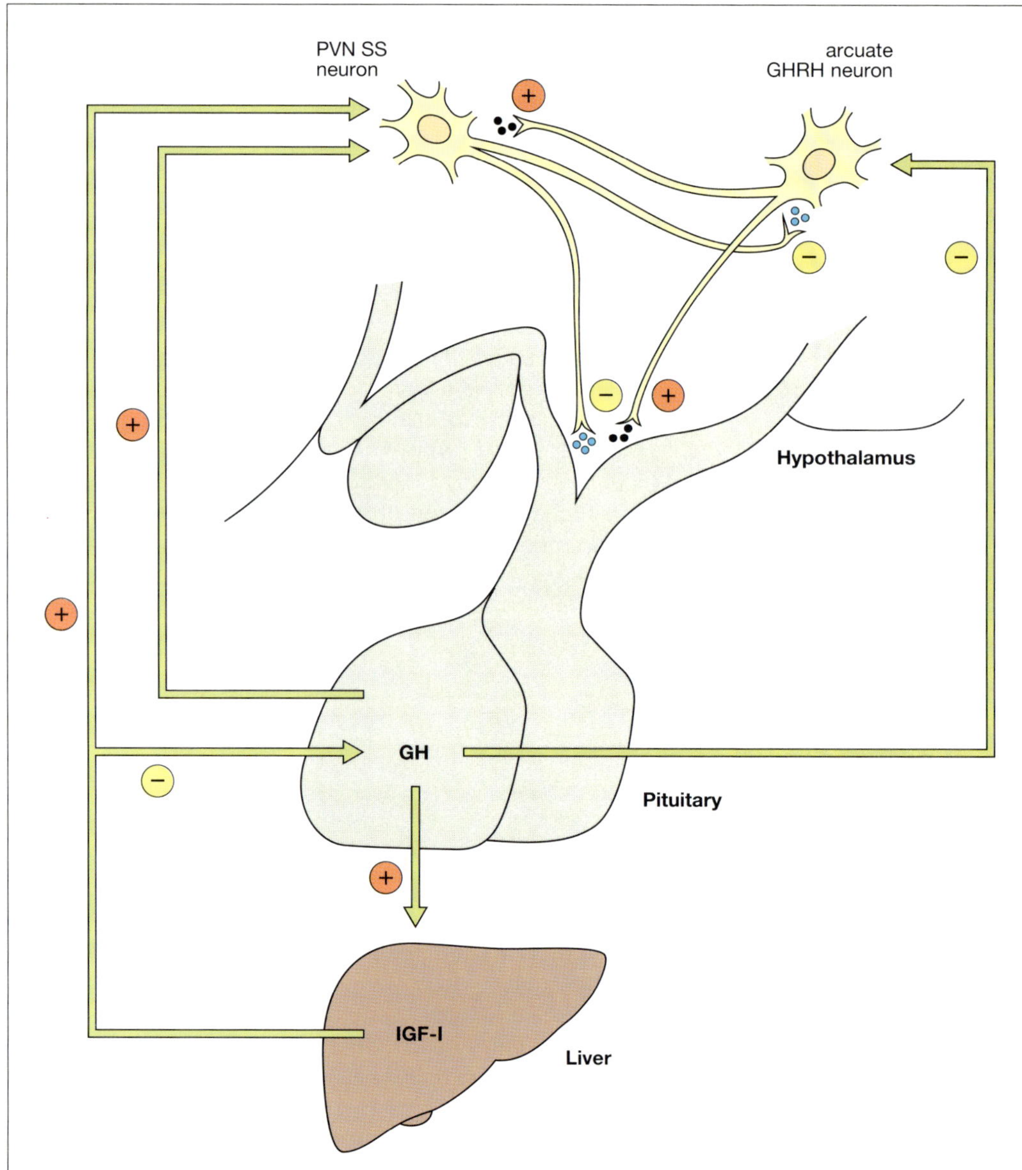

Fig. 10. *Schematic representation of the interaction between somatostatin **(SS)**, growth hormone releasing hormone **(GHRH)**, growth hormone **(GH)**, and insulin-like growth factor I **(IGF-I)** in regulating GH secretion. GH release is stimulated by GHRH (produced by GHRH neurons in the arcuate nucleus) and inhibited by SS (produced by somatostatinergic neurons in the anterior hypothalamic periventricular nucleus, PVN). SS inhibits GH secretion both by direct action at the pituitary level and indirectly through suppression of GHRH release. GHRH in turn stimulates SS secretion. GH exerts negative feedback on its own secretion by inhibiting GHRH release, stimulating SS release and potentiating the release of IGF-I from the liver. IGF-I in turn stimulates SS release and inhibits GH secretion by a direct action on the pituitary*

dence based on (a) the close anatomical proximity of SS cells to the glucagon-producing A cells and insulin-producing B cells, (b) the potent inhibitory effect of infused SS on insulin and glucagon secretion, and (c) the enhanced release of insulin and glucagon that occurs in some instances when islet D-cell secretion is suppressed argues in favour of a role of SS in modulating A- and B-cell secretion [30, 52, 79] (Fig. 11). Recent data on the islet microcirculation, however, suggest that the flow of blood within the islet is directed from the core to the periphery and occurs in an orderly manner from B to A to D cells and thence to the exterior of the islet [70]; such an arrangement does permit B- and A-cell secretion to reach and regulate D cells, but, because SS secreted from D cells is washed away from the islet, this peptide does not appear to play a significant role in modulating the release of insulin and glucagon through the islet vasculature [70]. The possibility that SS-14 and SS-28 may act differentially on islet cells is suggested by the finding that SS-14 is ten times more potent than SS-28 in inhibiting the secretion of glucagon whereas SS-28 exhibits tenfold higher activity than SS-14 in suppressing insulin secretion [40]. These findings correlate with the preferential expression of SS-14-type receptors on A cells and SS-28-type receptors on B cells [1, 48]. Mammalian D cells, however, produce predominantly SS-14, and in view of the negligible local islet production of SS-28 the question arises whether the SS-28-type receptors on B cells are vestigial or whether they interact with blood-borne SS-28 produced outside the pancreas. The latter possibility is clearly suggested by recent evidence showing that food ingestion stimulates the release of SS-28 from intestinal D cells into the circulation, where it achieves plasma concentrations sufficient to attenuate B-cell secretion and thereby qualify as a physiological modulator of nutrient-stimulated insulin release [14]. For these aspects and those dealt with in the next paragraph, see also the chapter by Schusdziarra, this volume.

The diffuse distribution of SS throughout the gut, together with its pleiotropic effects and the complex regulation of its secretion, suggests that SS exerts control over many discrete cell systems involved in GI, pancreatic and biliary functions such as absorption, secretion and motility [12, 22, 88, 90]. While the physiological basis of many of the gut actions of SS remains obscure, the recent development of useful models for the study of SS release and action has opened up parts of the GI tract, notably the stomach, to investigation. There is now evidence that circulating SS acts as a hormone to affect various GI functions [12, 22, 90]. SS regulates acid secretion both directly via the circulation to inhibit parietal cells and through a paracrine mechanism to suppress gastrin release [12, 39]. The presence of acid in the lumen of the stomach acts as a major stimulant of SS secretion and appears to be part of an inhibitory feedback mechanism for the regulation of gastric acid production. Circulating SS is also a physiological regulator of pancreatic exocrine secretion [22]. SS inhibits intestinal motility via modulation of the cAMP-dependent release of acetylcholine and cholinergic transmission. Elsewhere in the gut there is evidence to suggest that SS controls the rate of absorption of nutrients and participates in the regulation of gut hormone secretion, GI motor tone, blood flow and mucosal cell proliferation [12, 90]. Finally, there is good reason to believe that SS acting via the circulation interacts with adrenal glomerulosa cells to modulate angiotensin-stimulated aldosterone production.

Somatostatin in Disease

Despite the wide distribution of SS cells in the body, it is surprising that only a single disease – the somatostatinoma syndrome – has been attributed directly to SS dysfunction. Even in this condition, it is noteworthy that the considerable hypersomatostatinaemia is accompanied by relatively minor symptoms (gallstones, diarrhoea secondary to fat malabsorption and mild diabetes mellitus), a phenomenon probably due to tachyphylaxis and to the fact that many of the target cells on which SS normally acts locally are not accessible to circulating SS. Most somatostatinomas are actively secreting malignant islet-cell tumours that are associated with very high plasma levels of immunoreactive SS (600–15000 pmol/l) [31]. Lesser degrees of hypersomatostatinaemia have been observed in nonpancreatic SS-producing tumours such as duodenal somatostatinoma, extra-adrenal paraganglioma, pheochromocytoma and small-cell cancer of the lung. The serial measurement of plasma SS values has proved useful as a tumour marker in the follow-up of affected patients.

In a number of diseases, disordered SS function probably occurs as a secondary feature. Foremost among these is Alzheimer's disease, in which there is a decrease in the levels of SS in the cerebral cortex and CSF [4, 15]; this decrease is more severe in patients with a younger age of disease onset,

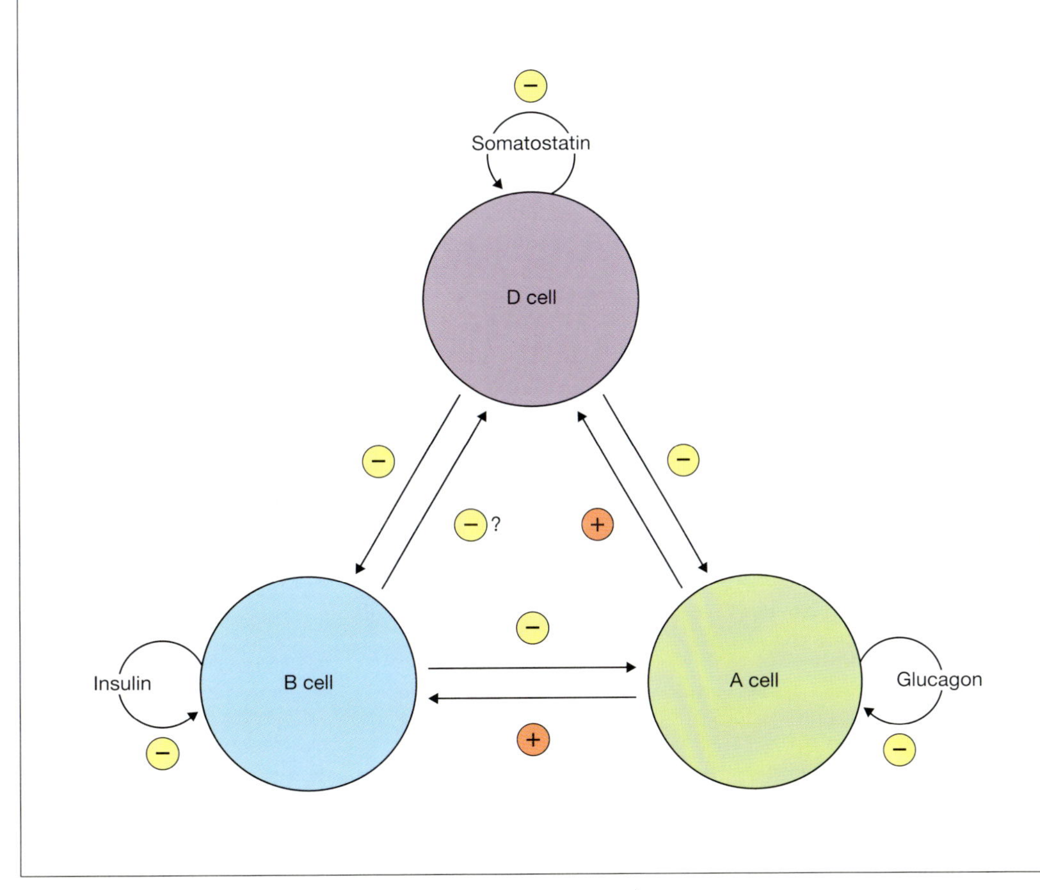

Fig. 11. *Effects of endogenous or exogenous SS, insulin and glucagon on the function of pancreatic islet cells. SS inhibits insulin and glucagon release, glucagon stimulates insulin and SS release, and insulin inhibits the release of glucagon and possibly of SS. In addition, all three islet hormones inhibit their own secretion by an autocrine mechanism. Physiologically, intraislet insulin and glucagon regulate the secretion of SS, and intraislet insulin regulates glucagon release. The precise physiological role of intraislet SS remains unclear (see text for details)*

and, although its pathophysiological significance is unclear, it has become an important biochemical marker for the disease. Cerebrocortical SS is also decreased in Parkinson's disease with dementia [4]. Significantly lower levels of SS in the CSF have been observed in depressed patients [69]. Huntington's disease is characterized by a loss of striatal neurons but selective preservation of a subclass of neurons that colocalize the enzyme NADPH diaphorase (NADPH-d), SS and NPY; in these preserved neurons, there are three- to fivefold increases in immunoreactive SS and NPY content [4]. Since in Huntington's disease the brain is capable of producing excessive quantities of the excitotoxin quinolinic acid, an NMDA-receptor agonist, and since quinolinic acid experimentally induces neuronal loss with sparing of NADPH-d/SS/NPY neurons in vivo as well as increased production of SS peptide and mRNA levels by cultured brain cells, it has been suggested that an NMDA-receptor-mediated effect may be involved in the genesis of the changes affecting SS in Huntington's disease and perhaps in the pathophysiology of the condition [4, 46]. Alterations in hippocampal SS cells accompanied by a reduction of SS levels in the CSF have been observed in both experimental and human epilepsy. These changes, however, may be nonspecific since other peptides such as NPY, neurotensin, cholecystokinin (CCK) and vasoactive intestinal peptide (VIP) are similarly altered [67]. For more information on SS in neuropsychiatric disorders, see the chapter by Rubinow et al. (this volume).

Plasma SS levels are elevated significantly in hepatic cirrhosis and in chronic renal failure as a result of the peptide's impaired metabolism [76]. Despite the large quantities of SS present in the gut, primary GI disease is generally not associated with alterations in the concentrations of circulating SS. In experimental insulinopenic diabetes, there is marked hypersomatostatinaemia secondary to SS hypersecretion by pancreatic islet and gastric D cells in response to hypoinsulinaemia [19]; this change is readily reversed by insulin administration. However, the situation in patients with type I diabetes mellitus has not been adequately documented. In experimental hyperinsulinaemic diabetes and in human type II diabetes there is an impaired release of gut SS in response to the ingestion of a meal.

References

1. Amherdt M, Patel YC, Orci L (1987) Selective binding of somatostatin-14 and somatostatin-28 to islet cells revealed by quantitative electron microscopic autoradiography. J Clin Invest 80: 1455–1458
2. Argos P, Taylor WL, Minth CD, Dixon JE (1983) Nucleotide and amino acid sequence comparisons of preprosomatostatins. J Biol Chem 258: 8788–8793
3. Baetens D, Malaisse-Lagae F, Perrelet A, Orci L (1979) Endocrine pancreas: three dimensional reconstruction shows two types of islets of Langerhans. Science 206: 1323–1325
4. Beal MF (1990) Somatostatin in neurogenerative illnesses. Metabolism [Suppl 2] 39: 116–119
5. Benoit R, Ling N, Esch F (1987) A new prosomatostatin derived peptide reveals a pattern for prohormone cleavage at monobasic sites. Science 238: 1126–1129
6. Berelowitz M, Dudlak D, Frohman LA (1982) Release of somatostatin-like immunoreactivity from incubated rat hypothalamus and cerebral cortex. J Clin Invest 69: 1293–1301
7. Berelowitz M, Firestone S, Frohman LA (1981) Effects of growth hormone excess and deficiency on hypothalamic somatostatin content and release and on tissue somatostatin distribution. Endocrinology 109: 714–719
8. Berelowitz M, Maeda K, Harris S, Frohman LA (1980) The effect of alterations in the pituitary-thyroid axis on hypothalamic content and in vitro release of somatostatin-like immunoreactivity. Endocrinology 107: 24–29
9. Berelowitz M, Szabo M, Frohman LA, Firestone S, Chu L (1981) Somatomedin-C mediates growth hormone negative feedback by effects on both the hypothalamus and the pituitary. Science 212: 1279–1281
10. Brazeau P, Vale WW, Burgus R, Ling N, Butcher M, Rivier J, Guillemin R (1973) Hypothalamic peptide that inhibits the secretion of immunoreactive pituitary growth hormone. Science 179: 77–79
11. Chowen-Breed J, Steiner RA, Clifton DK (1989) Sexual dimorphism and testosterone dependent regulation of somatostatin gene expression in the periventricular nucleus of the rat brain. Endocrinology 125: 357–362
12. Colturi TJ, Unger RH, Feldman M (1984) Role of circulating somatostatin in regulation of gastric acid secretion, gastrin release and islet cell function. Studies in healthy subjects and duodenal ulcer patients. J Clin Invest 74: 417–423
13. Critchlow V, Abe S, Urman S, Vale WW (1981) Effects of lesions of the periventricular nucleus of the preoptic-anterior hypothalamus on growth hormone and thyrotropin secretion and brain somatostatin. Brain Res 222: 267–276
14. D'Alessio DA, Sieber C, Beglinger C, Ensinck JW (1989) A physiologic role for somatostatin-28 as a regulator of insulin secretion. J Clin Invest 84: 857–862
15. Davies P, Katzman R, Terry RD (1980) Reduced somatostatin like immunoreactivity in cerebral cortex from cases of Alzheimer's disease and Alzheimer senile dementia. Nature 288: 279–280
16. Epelbaum J (1986) Somatostatin in the central nervous system: physiology and pathological modifications. Prog Neurobiol 27: 63–100
17. Finley JCW, Maderdrut JL, Roger LJ, Petrusz P (1981) The immunocytochemical localization of somatostatin-containing neurons in the rat central nervous system. Neuroscience 6: 2173–2192
18. Funckes CL, Minth CD, Deschenes R, Magazin M, Tavianini MA, Sheets M, Collier K, Weith HL, Aron DL, Roos BA, Dixon JE (1983) Cloning and characterization of a mRNA encoding rat preprosomatostatin. J Biol Chem 258: 8781–8787
19. Gerich JE (1990) Role of somatostatin and its analogues in the pathogenesis and treatment of diabetes mellitus. Metabolism [Suppl 2] 39: 52–54
20. Goodman RH, Aron DC, Roos BA (1982) Rat preprosomatostatin: structure and processing by microsomal membranes. J Biol Chem 258: 5570–5573
21. Goodman RH, Rehfuss RP, Verhave M, Ventimiglia R, Low MJ (1990) Somatostatin gene regulation: an overview. Metabolism [Suppl 2] 39: 2–5
22. Gyr K, Beglinger C, Kohler E, Trantzl U, Keller U, Bloom SR (1987) Circulating somatostatin: physiological regulator of pancreatic function? J Clin Invest 79: 1595–1600
23. Hellman B, Lernmark A (1969) Inhibition of the in vitro secretion of insulin by an extract of pancreatic α_1 cells. Endocrinology 84: 1484–1487
24. Hobart P, Crawford R, Shen L-P, Pictet R, Rutter WJ (1980) Cloning and sequence analysis of cDNAs encoding two distinct somatostatin precursors found in the endocrine pancreas of anglerfish. Nature 288: 137–141
25. Horvath S, Palkovits M, Gorcs T, Arimura A (1989) Electron microscopic immunocytochemical evidence for the existence of bidirectional synaptic connections between growth hormone releasing hormone- and somatostatin-containing neurons in the hypothalamus of the rat. Brain Res 481: 8–15

26. Ikeda SR, Schofield GG (1989) Somatostatin blocks a calcium current in rat sympathetic ganglion neurones. J Physiol (Lond) 409: 221–240
27. Johansson O, Hokfelt T, Elde RP (1984) Immunohistochemical distribution of somatostatin-like immunoreactivity in the central nervous system of the adult rat. Neuroscience 13: 265–339
28. Katakami H, Arimura A, Frohman LA (1986) Growth hormone (GH)-releasing factor stimulates hypothalamic somatostatin release: an inhibitory feedback effect on GH secretion. Endocrinology 118: 1872–1877
29. Katakami H, Downs TR, Frohman LA (1988) Inhibitory effect of hypothalamic medial preoptic area somatostatin on growth hormone releasing factor in the rat. Endocrinology 123: 1103–1109
30. Kawai K, Orci L, Ipp E, Perrelet A, Unger RH (1982) Circulating somatostatin acts on the islets of Langerhans via a somatostatin-poor compartment. Science 218: 477–478
31. Krejs GJ, Orci L, Conlon JM, Ravazzola M, Davis GR, Ruskin P, Collins SM, McCarthy DM, Baetens D, Rubenstein A, Aldor TAM, Unger RH (1979) Somatostatinoma syndrome: biochemical, morphological and clinical features. N Engl J Med 301: 285–292
32. Krisch B (1979) Immunohistochemical results on the distribution of somatostatin in the hypothalamus and in limbic structures of the rat. J Histochem Cytochem 27: 1389–1390
33. Krulich L, Dhariwal APS, McCann SM (1968) Stimulatory and inhibitory effects of purified hypothalamic extracts on growth hormone release from rat pituitary in vitro. Endocrinology 83: 783–790
34. Larsson LI (1985) Distribution and morphology of somatostatin cells. Adv Exp Biol Med 188: 383–402
35. Liebow C, Reilly C, Serrano M, et al. (1989) Somatostatin analogues inhibit growth of pancreatic cancer by stimulating tyrosine phosphatase. Proc Natl Acad Sci USA 86: 2003–2007
36. Luini A, de Matteis MA (1990) Evidence that receptor-linked G protein inhibits exocytosis by a post-second messenger mechanism in AtT-20 cells. J Neurochem 54: 30–38
37. Mackin RB, Noe BD (1987) Direct evidence for two distinct prosomatostatin converting enzymes. J Biol Chem 262: 6453–6456
38. Magazin M, Minth CD, Funckes CL, Deschenes R, Tavianini MA, Dixon JE (1982) Sequence of a cDNA encoding pancreatic preprosomatostatin-22. Proc Natl Acad Sci USA 79: 5152–5156
39. Makhlouf GM, Shubert ML (1990) Gastric somatostatin: a paracrine regulator of acid secretion. Metabolism [Suppl 2] 39: 138–142
40. Mandarino L, Stenner D, Blanchard W, Nissen S, Gerich J, Ling N, Brazeau P, Bohlen P, Esch F, Guillemin R (1981) Selective effects of somatostatin-14, -25, and -28 on in vitro insulin and glucagon secretion. Nature 291: 76–77
41. McDonald JK, Greiner F, Bauer GE, Elde RP, Noe BD (1987) Separate cell types that express two different forms of somatostatin in anglerfish islets can be immunohistochemically differentiated. J Histochem Cytochem 35: 155–162
42. Montminy MR, Goodman RH, Horovitch SJ, Habener JF (1984) Primary structure of the gene encoding rat preprosomatostatin. Proc Natl Acad Sci USA 81: 3337–3340
43. Montminy MR, Gonzalez GA, Yamamoto KK (1990) Characteristics of the cAMP response unit. Metabolism [Suppl 2] 39: 6–12
44. Noe BD, Debo G, Spiess J (1984) Comparison of prohormone processing activities in islet microsomes and secretory granules: evidence for distinct converting enzymes for separate islet prosomatostatin. J Cell Biol 99: 578–587
45. Papachristou DN, Patel YC (1988) Glucocorticoid (GL) regulation of somatostatin (S) gene expression in normal tissues and in a S-producing tumor cell line (1027 B_2). In: Program Annual Meeting Society For Neuroscience, Toronto, Canada, vol 14, part I, p 17 (Abstr 11.5)
46. Patel SC, Papachristou DN, Patel YC (1991) Quinolinic acid stimulates somatostatin gene expression in cultured rat cortical neurons. J Neurochem 56: 1286–1291
47. Patel YC (1990) Somatostatin. In: Becker K (ed) Principles and practice of endocrinology and metabolism. Lippincott, Philadelphia, pp 1297–1301
48. Patel YC, Amherdt M, Orci L (1982) Quantitative electron microscopic radiography of insulin, glucagon and somatostatin binding sites on islets. Science 217: 1155–1156
49. Patel YC, Murthy KK, Escher E, Banville D, Spiess J, Srikant CB (1990) Mechanism of action of somatostatin: an overview of receptor function and studies of the molecular characterization and purification of somatostatin receptor proteins. Metabolism [Suppl 2] 39: 63–69
50. Patel YC, O'Neil W (1988) Peptides derived from cleavage of prosomatostatin at carboxy and amino terminal segments: characterization of tissue and secreted forms in the rat. J Biol Chem 263: 745–751
51. Patel YC, Papachristou DN, Zingg HH, Farkas EM (1991) Regulation of islet somatostatin secretion and gene transcription: selective effects of cAMP and phorbol esters in normal islets and in a somatostatin producing rat islet clonal cell line (1027 B_2). Endocrinology 128: 1754–1762
52. Patel YC, Pierzchala I, Amherdt M, Orci L (1985) Effects of cysteamine and antibody to somatostatin on islet cell function in vitro: evidence that intracellular somatostatin deficiency augments insulin and glucagon secretion. J Clin Invest 75: 1249–1255
53. Patel YC, Prakash MLS, Papachristou DN, Farkas EM, Pham K (1988) Insulin is a potent inhibitor of somatostatin secretion and mRNA accumulation in a somatostatin-producing islet tumor cell line. Diabetes [Suppl 1] 37: 101 A
54. Patel YC, Rao K, Reichlin S (1977) Somatostatin in human cerebrospinal fluid. N Engl J Med 296: 529–533
55. Patel YC, Reichlin S (1978) Somatostatin in hypothalamus, extrahypothalamic brain and peripheral tissues of the rat. Endocrinology 102: 523–530
56. Patel YC, Srikant CB (1985) Somatostatin mediation of adenohypophysial secretion. Annu Rev Physiol 48: 551–567
57. Patel YC, Tannenbaum GS (eds) (1985) Somatostatin. Adv Exp Med Biol 188
58. Patel YC, Tannenbaum GS (eds) (1990) Somatostatin. Basic and clinical aspects. Metabolism [Suppl 2] 39
59. Patel YC, Wheatley T, Ning C (1981) Multiple forms of immunoreactive somatostatin: comparison of distribution in neural and nonneural tissues and portal plasma of the rat. Endocrinology 109: 1943–1949
60. Peterfreund RA, Vale WW (1985) Somatostatin secretion from the hypothalamus. Adv Exp Med Biol 188: 183–200
61. Plotsky PM, Vale WW (1985) Patterns of growth hormone releasing factor and somatostatin secretion into hypophysial portal circulation of the rat. Science 230: 461–463
62. Rabbani SN, Patel YC (1988) Measurement and characterization of somatostatin-14 like immunoreactivity in human urine. J Clin Endocrinol Metab 66: 1050–1055
63. Rabbani SN, Patel YC (1990) Peptides derived by processing of rat prosomatostatin near the amino terminus: characterization, tissue distribution and release. Endocrinology 126: 2054–2061
64. Reichlin S (1983) Somatostatin. N Engl J Med 309: 1495–1501, 1556–1563
65. Reichlin S (ed) (1987) Somatostatin basic and clinical status. Plenum, New York
66. Reubi JC, Kvols L, Krenning E, Lamberts SWJ (1990) Distribution of somatostatin receptors in normal and tumor tissue. Metabolism [Suppl 2] 39: 78–81
67. Riekkinen P, Pitkanen A (1990) Somatostatin and epilepsy. Metabolism [Suppl 2] 39: 112–115
68. Rogers KV, Vician L, Steiner RA, Clifton DK (1988) The effect of hypophysectomy and growth hormone administration on preprosomatostatin messenger ribonucleic acid in the periventricular nucleus of the rat hypothalamus. Endocrinology 122: 586–590
69. Rubinow DR, Gold PW, Post RM, Ballenger JC, Cowdry R, Rollinger J, Reichlin S (1983) CSF somatostatin in affective illness. Arch Gen Psychiatry 40: 409–412
70. Samols E, Stagner JI (1990) Islet somatostatin – microvascular, paracrine and pulsatile regulation. Metabolism [Suppl 2] 39: 55–60
71. Sasaki A, Yoshinaga K (1989) Immunoreactive somatostatin in male reproductive system in humans. J Clin Endocrinol Metab 68: 996–999
72. Scarborough DE (1990) Somatostatin regulation by cytokines. Metabolism [Suppl 2] 39: 108–111

73. Schonbrunn A (1990) Somatostatin action in pituitary cells involves two independent transduction mechanisms. Metabolism [Suppl 2] 39: 96–100
74. Sevarino KA, Stork P, Ventimeglia R, Mandel G, Goodman RH (1989) Processing and intracellular sorting of anglerfish and rat preprosomatostatins in mammalian endocrine cells. Cell 57: 11–19
75. Shen L-P, Pictet RL, Rutter WJ (1982) Human somatostatin. I. Sequence of the cDNA. Proc Natl Acad Sci USA 79: 4575–4579
76. Shoelson SE, Polonsky KS, Nakabayashi T, Jaspan JB, Tager HS (1986) Circulating forms of somatostatin-like immunoreactivity in human plasma. Am J Physiol 250: E428–E434
77. Skamene A, Patel YC (1984) Infusions of graded concentrations of somatostatin-14 in man: pharmacokinetics and differential inhibitory effects on pituitary and islet hormones. Clin Endocrinol (Oxf) 20: 555–564
78. Srikant CB, Patel YC (1987) Somatostatin receptors: evidence for functional and structural heterogeneity. In: Reichlin S (ed) Proceedings of the International Conference on Somatostatin. Plenum, New York, pp 89–102
79. Taborsky GJ (1983) Evidence of a paracrine role for pancreatic somatostatin in vivo. Am J Physiol 245: E598–E603
80. Tannenbaum GS, McCarthy GF, Zeitler P, Beaudet A (1990) Cysteamine induced enhancement of growth hormone releasing factor (GRF) immunoreactivity in arcuate neurons: morphological evidence for putative somatostatin/GRF interactions within hypothalamus. Endocrinology 127: 2551–2560
81. Tannenbaum GS, Painson J-C, Lapointe M, Gurd W, McCarthy GF (1990) Interplay of somatostatin and growth hormone releasing hormone in genesis of episodic growth hormone secretion. Metabolism [Suppl 2] 39: 35–39
82. Taylor WL, Collier KJ, Deschenes RJ, Weith HL, Dixon JE (1981) Sequence analysis of a cDNA coding for a pancreatic precursor to somatostatin. Proc Natl Acad Sci USA 78: 6694–6698
83. Thorner MO, Vance ML, Hartman ML, Hall RW, Evans WS, Veldhuis JD, van Cauter E, Copinschi G, Bowers CY (1990) Physiological role of somatostatin on growth hormone regulation in humans. Metabolism [Suppl 2] 39: 40–42
84. Wang H-L, Bogen C, Reisine T, Dichter M (1989) Somatostatin-14 and somatostatin-28 induce opposite effects on potassium currents in rat neocortical neurons. Proc Natl Acad Sci USA 86: 9616–9620
85. Weir GC, Bonner-Weir S (1985) Pancreatic somatostatin. Adv Exp Med Biol 188: 403–423
86. Werner H, Koch Y, Baldino F Jr, Gozes I (1988) Steroid regulation of somatostatin mRNA in the rat hypothalamus. J Biol Chem 263: 7666–7674
87. Wollheim CB, Winiger BP, Ullrich S, et al. (1990) Somatostatin inhibition of hormone release: effects on cytosolic Ca^{++} and interference with distal secretory events. Metabolism [Suppl 1] 39: 101–104
88. Yamada T (1987) Gut somatostatin. In: Reichlin S (ed) Somatostatin. Basic and clinical status. Plenum, New York, pp 221–228
89. Yatani A, Codina J, Sekura RD, et al. (1987) Reconstitution of somatostatin and muscarinic receptor mediated stimulation of K^+ channels by isolated G_k protein in clonal rat anterior pituitary cell membranes. Mol Endocrinol 1: 283–289
90. Zyznar E, Pietri A, Harris V, Unger RH (1981) Evidence for the hormonal status of somatostatin in man. Diabetes 30: 883–886

Somatostatin Receptors in the Central Nervous System

J. Epelbaum

U 159 INSERM, 2ter rue d'Alésia, 75014 Paris, France

Introduction

Somatostatin (SS), a peptide of 14 amino acids, is a hydrophobic molecule. Its multiple physiological effects derive from its binding to an outer membrane receptor. This chapter will focus on the characterization of SS binding sites, their coupling with multiple second messenger systems, their localization and their multiple functions. Finally, we shall briefly review the evidence for the possible involvement of SS receptors in brain pathology.

Somatostatin Receptors

In Vitro Binding Assays

Since the original study of Schonbrunn and Tashjian [94] on a clonal pituitary cell line (GH_4C_1 cells), numerous authors have described the properties of nanomolar-affinity SS binding sites in the brain [8, 11, 19, 37, 70, 98] and pituitary [1, 16, 38, 82, 100]. In saturation experiments in both tissues, various monoiodinated tyrosine-substituted SS agonists bind to a single class of binding sites. Surprisingly, it was also found that SS had a weak but measurable affinity for opioid receptors [105].

Step-by-step modification of the ring structure of SS made it possible to pinpoint the active site of the molecule to amino acids 7–10 and to design shorter-ringed analogues less prone to enzymatic degradation (see Fig. 1). In the brain cortex, hippocampus and striatum, some of these shorter analogues such as octreotide displaced ^{125}I-SS binding in a biphasic manner [73, 107], whereas the displacement curves were monophasic in other target organs, e.g. the pituitary and the pancreas (Fig. 2). Thus, while iodinated ligands bind to SS binding sites with the same affinity in all target organs, octreotide seems to be able to differentiate between two classes of sites in the brain, one with nanomolar (SSA or SS1) and the other with micromolar (SSB or SS2) affinity for the peptide. However, it should be kept in mind that octreotide is quite a good μ-opioid antagonist, only 20 times less potent than naloxone on a molar basis [54]; furthermore the most selective and potent μ-opioid antagonists yet synthesized are a series of six octapeptides structurally related to SS [43].

Two peptides of the SS family are bioactive and present in the brain: SS-14 and its N-terminal extension SS-28. They derive from the same precursor molecule and, in most instances, are located in the same nerve terminals, from which they are coreleased in a calcium-dependent manner (reviews in [17, 71]). Only one set of neurons in the brain stem has been shown to be immunopositive for SS-28 alone [10]. With in vitro binding assays it has not been possible to describe pharmacologically distinct binding sites for SS-28 and SS-14. SS-28 has a slightly higher affinity than SS-14, but both peptides are full agonists in every binding assay tested to date [47, 48, 60, 90, 99].

In the brain, as in the pituitary, ^{125}I-SS binding is regulated positively by divalent cations [62, 80] and negatively by guanine nucleotides [15, 60]. This is strongly in favour of the coupling of SS binding sites with guanine-nucleotide- and magnesium-binding proteins (G-proteins), which are involved in transducing systems mediated by cyclic adenosine monophosphate (cAMP) and/or calcium. In such systems, the interaction of guanosine triphosphate (GTP) with the G-protein increases the coupling to the effector and decreases the affinity of the agonist to the receptor (Fig. 3).

Structural Characterization

Since 1986, several attempts have been made at purifying SS receptors, either from peripheral organs like the pancreas [39, 89, 103, 104], the adrenal cortex [102], the stomach [84, 85] and the pituitary [6, 37, 50, 63] or from the brain [27, 28, 37, 90, 106]. Different techniques (chemical cross-linking, affinity and photoaffinity labelling and various purification procedures) were used in these studies, yielding mainly contradictory data (see Table 1). For instance, using ^{125}I-Tyr$_{11}$-SS as a ligand, Murthy et al. [63] purified three binding sites of 57, 42 and 27 kDa from pituitary cell lines, whereas with the same ligand but on normal pituitary cell preparations Lewis and Williams [50] found a single glycoprotein of 88 kDa and Bruno and Berelowitz [6] obtained two binding sites of 69 and 66 kDa. On the other hand, when brain and pituitary receptors were purified in the same study, either the same [39, 106] of different [90] molecular weights were reported. No definitive agreement has thus been reached on

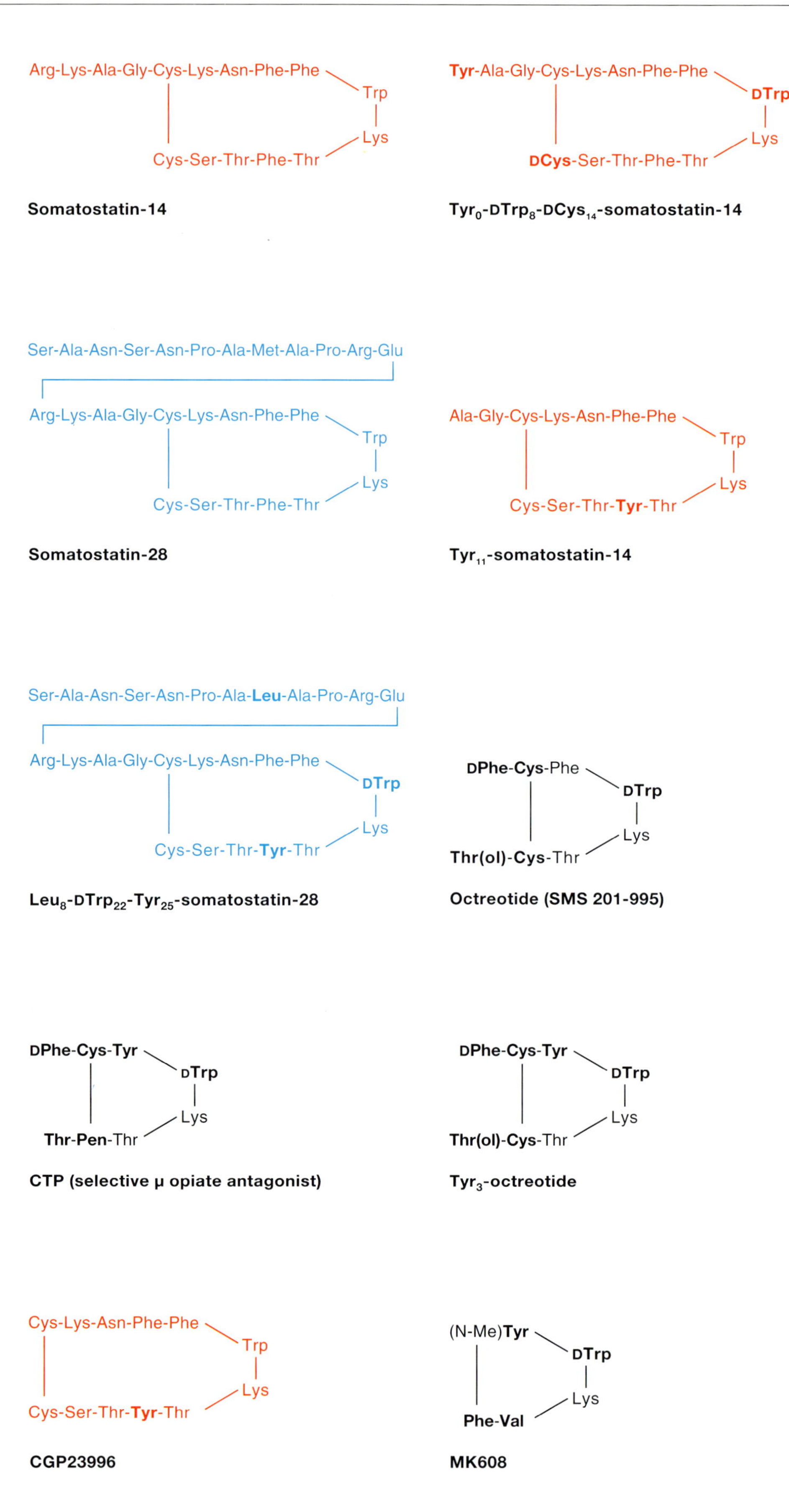

Fig. 1. *Chemical structure of several SS-receptor agonists derived from SS-14* ***(red)****, SS-28* ***(blue)*** *and octreotide* ***(black)****. The modified amino acids are printed in* ***bold.*** *Agonists mono-substituted with a tyrosine residue are used as radioactive ligands after labelling with* ^{125}*Iodine. In the structure of CTP,* ***Pen*** *stands for penicillamine*

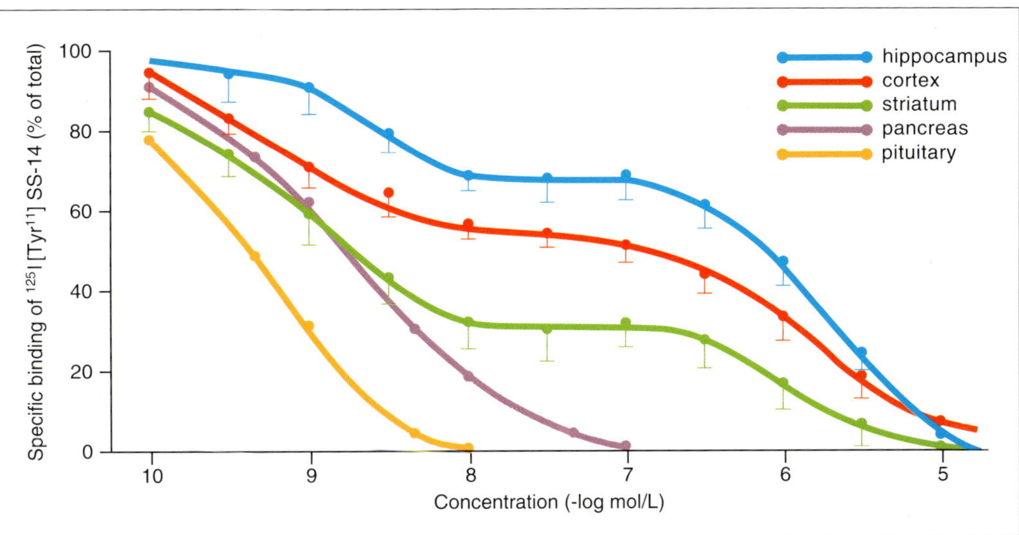

Fig. 2. *Competition between* ^{125}*I-Tyr*$_{11}$*-SS-14 and octreotide showing regional and organ variation of the relative proportions of high-affinity (SSA) and low-affinity (SSB) SS binding sites. The SSA receptors represent 45% of the binding in the cortex (red curve), 30% in the hippocampus (blue curve), 70% in the striatum (green curve) and 100% in the pituitary (orange curve) and pancreas (purple curve). Data points are means of five separate estimations.* (Modified from [107])

Table 1. *Characteristics of SS receptors*

Tissue and source	*Molecular mass (kDa)*	*Methods*	*Ligands*[a]	*Properties*
Brain	70	Cross-linking	Tyr$_{11}$-SS-14, Leu$_8$-DTrp$_{22}$--Tyr$_{25}$-SS-28	GTP, WGA
	60	Cross-linking	CGP23996	GTP, WGA
	71	Cross-linking	Tyr$_{11}$-SS-14	
	60	Solubilization, affinity chromatography	CGP23996	
	400	Solubilization, immunoprecipitation with G_i antibodies	MK608	GTP, WGA
Pituitary	88	Cross-linking	Tyr$_{11}$-SS-14	GTP, WGA
	94	Cross-linking	Tyr$_{11}$-SS-14	GTP, WGA
	82			Mg^{2+}, E2
	69, 66, 45	Cross-linking	Tyr$_{11}$-SS-14	
	60	Cross-linking	CGP23996	GTP, WGA
	60	Cross-linking	MK608	GTP, WGA
AtT20	55	Cross-linking	CGP23996	
AtT20 and GH3	57	Cross-linking	Tyr$_{11}$-SS-14, Leu$_8$-DTrp$_{22}$--Tyr$_{25}$-SS-28, Tyr$_3$-SMS	
Pancreas	92	Cross-linking	Tyr$_{11}$-SS-14	GTP, WGA
	200, 80, 70	Cross-linking	Tyr$_{11}$-SS-14	
	94	Cross-linking	Leu$_8$-DTrp$_{22}$--Tyr$_{25}$-SS-28	
	400 100, 56, 21	Solubilization, polyacrylamide gel electrophoresis	Tyr$_3$-SMS	GTP
Hamster insulinoma	193, 129, 42	Cross-linking	Leu$_8$-DTrp$_{22}$--Tyr$_{25}$-SS-28	
Gastric tumoural cell line	90	Solubilization	Tyr$_{11}$-SS-14	
Adrenal cortex	200	Cross-linking	Tyr$_{11}$-SS-14, Leu$_8$-DTrp$_{22}$--Tyr$_{25}$-SS-28	

[a] The binding is sensitive to guanine nucleotides (GTP), divalent cations such as magnesium (Mg^{2+}) and/or oestradiol (E2). WGA: the purified putative receptor protein is retained on a chromatographic wheat-germ-agglutinin column; this indicates that the receptor is a glycoprotein, a necessary condition for a membrane-bound receptor; SMS, octreotide

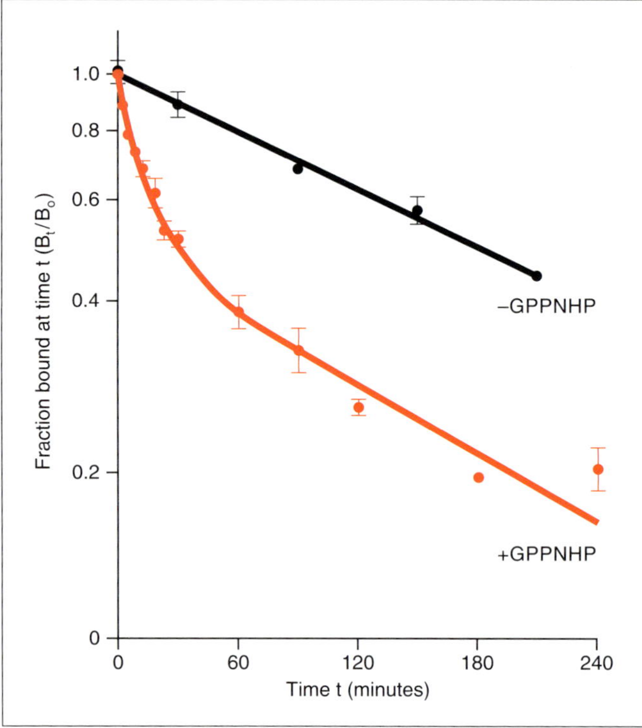

Fig. 3. *Increase in the dissociation rate of* ^{125}*-I-Tyr*$_1$*-SS-14 binding in the presence of guanine nucleotide. The amount of* ^{125}*I-Tyr*$_1$*-SS-14 bound was determined in the presence (red curve) or absence of 0.3 μmol/l guanylyl imidodiphosphate (GPPNHP). Data are mean ± SE from triplicate samples. (Modified from [93])*

the SS receptor size, and it remains to be demonstrated that there are different SS receptor molecules. However, most authors agree that the migration of the labelled proteins is not modified by thiol reducing agents, a fact indicating that there is a general lack of disulfide bridges in the SS receptor. The binding of the solubilized sites to wheat germ agglutinin seen in most studies indicates that the protein is glycosylated, and differing glycosylation patterns could account for the diversity of the molecular weights in different tissues. Finally, the guanine nucleotide dependency of the binding is consistent with the coupling of the purified SS binding sites to GTP-binding proteins.

Using a different approach, one study has recently reported the functional expression of pituitary SS receptors in *xenopus* oocytes after the injection of RNA isolated from the anterior pituitary cell line AtT20 [115]. This technique should provide a simpler procedure than mere purification for the determination of the structure of the SS receptors, i.e. cDNA cloning. Furthermore, it would enable the putative subtypes to be purified according to their different effector coupling mechanisms (see "Mechanisms of Action").

Localization of Somatostatin Receptors by Radioautography

In the brain, the localization of a given receptor is an important clue to the function of its endogenous ligand. In the 1970s, Kuhar and Yamamura [44] developed an in vitro radioautographic technique making the same type of experiment possible as with the classical membrane binding technique, but directly on cryostat-cut brain sections.

Film radioautography (Fig. 4) after incubation with various iodinated or tritiated ligands [20, 26, 46–48, 55, 56, 65, 70, 74, 79, 81, 83, 109, 116] has shown SS binding sites to be widely distributed in mammalian brains. They are highly concentrated in the deeper cortical layers (V and VI) throughout the neuraxis, most regions of the limbic system such as the hippocampus (molecular and granular layers of the dentate gyrus, all but the pyramidal layer of CA1), the amygdala (mainly the basolateral nucleus), the septum and the habenula. Regions involved in sensory functions are also rich in SS receptors, e.g. the olfactory tubercles, the anterior olfactory nucleus, the inferior and superior colliculi and the retina. Moderate levels are present in the caudate-putamen and the thalamus. The cerebellum is totally devoid of binding sites in adult rat brain. In the hypothalamus, the first studies reported only few binding sites in anterior regions, but preincubation of the sections with desaturating concentrations of GTP [46] revealed an important proportion of SS binding sites occupied by the peptide. In summary, there is a general correlation between the regional levels of SS and the numbers of receptors, with a high rate of binding site occupancy. The only discrepancy in the literature concerns the localization of SS receptors in the substantia nigra, since they have been visualized in some studies [26, 48] but not in others [20, 79, 109, 116]. According to a recent report, SS is able to stimulate cAMP formation in this region [53], whereas the peptide is usually inhibitory in other brain regions and in the pituitary (see "Mechanisms of Action").

By means of light microscopic resolution radioautography, it was observed that SS receptors are associated with specific cell groups in a number of brain regions. At high magnification, the label in the medial habenula was seen to predominate over zones of densely packed neuronal perikarya that correspond to the cholinergic cell group (Ch7) projecting to the interpeduncular nucleus. One of the most important adrenergic cell groups, the locus coeruleus, has been shown to contain the highest SS binding levels of the entire neuraxis. The labelling was apparent throughout the rostrocaudal extent of the nucleus in close association with tyrosine hydroxylase-immunoreactive neurons detected on adjacent sections. In contrast to the homogeneous distribution observed in the medial habenula and the locus coeruleus, a cluster of ^{125}I-SS-labelled cells was detected throughout the rostrocaudal extent of the arcuate nucleus [18]. These cells were regrouped in the ventrolateral aspect of the nucleus as well as around the mammillary recess of the third ventricle, and their distribution corresponded to the immunohistochemical localization of growth-hormone-releasing factor (GHRF) neurons; this finding provided an anatomical substrate to the intrahypothalamic interactions

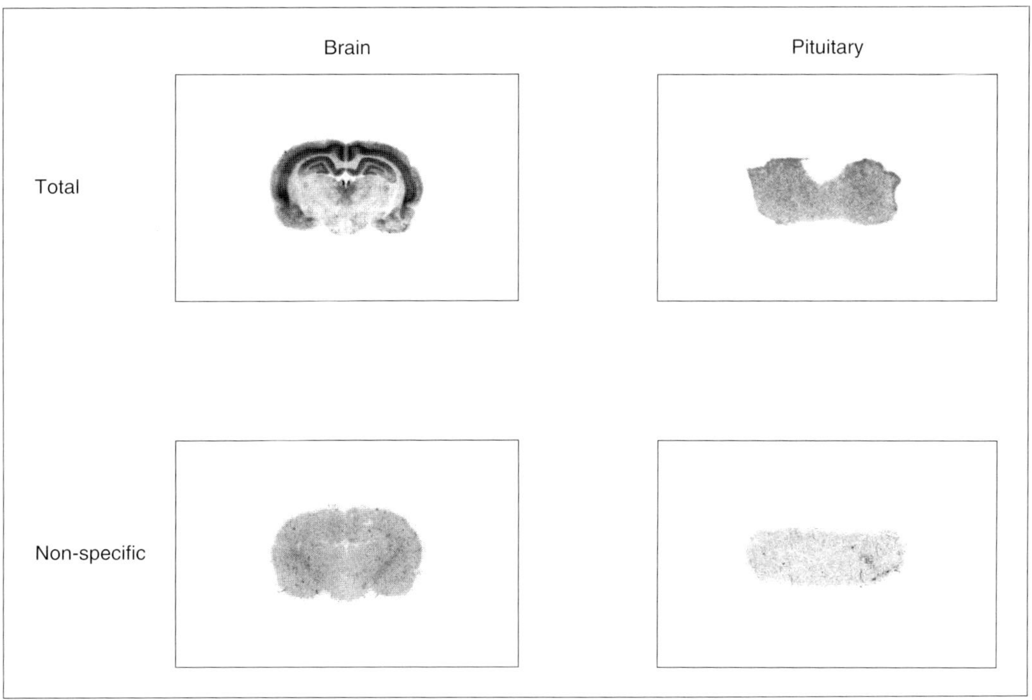

Fig. 4. *Radioautographic localization of* ^{125}*-I-Tyr*$_0$*-*D*Trp*$_8$*-SS-14 binding sites in coronal sections from rat brain and pituitary. Frozen sections of rat brain and pituitary tissues were incubated in the presence of* ^{125}*I-Tyr*$_0$*-*D*Trp*$_8$*-SS-14 0.25 nmol/l, in the absence (total binding) or presence (non-specific binding) of SS-14 0.1 µmol/l*

between SS and GHRF that have been postulated in the control of GH secretion.

In the pituitary, SS receptors are located only in the anterior lobe. At the ultrastructural level, after internalization of the ligand the binding is located in somatotrophs, thyrotrophs and lactotrophs [59]. Curiously, internalization has not been observed on clonal pituitary cell lines [67].

Mechanisms of Action

SS receptors seem to be able to recruit three types of intracellular mechanisms to transduce the various effects of the peptide: (1) inhibition of cAMP formation, (2) diminution of intracellular calcium concentrations and (3) stimulation of a phosphoprotein phosphatase.

As early as 1980, it was found that micromolar quantities of SS were able to inhibit the formation of cAMP in brain cell [111] and pituitary membrane [88] preparations. Resolution of brain culture in neuronal and glial preparations enabled SS-induced inhibition of adenylate cyclase activity to be observed in both cell types [9]. The agonist potency of SS analogues on the inhibition of adenylate cyclase activity was strictly correlated with their binding affinities [8]. The same order of activity of agonists on the inhibition of adenylate cyclase and of hormone secretions was observed in normal pituitary cell cultures [21]. However, the very marked differences between the relative affinities of SS for its binding sites and its inhibition of hormonal secretions on the one hand and adenylate cyclase on the other raised the question of a direct coupling of the SS receptor to this unique transducing mechanism. It was further found [21, 30, 40–42, 117] that SS could block equally hormonal secretions stimulated either by the cAMP pathway – stimulation of growth hormone (GH) secretion by growth-hormone-releasing hormone (GHRH) or of prolactin (PRL) secretion by vasoactive intestinal peptide VIP – or through cAMP-independent pathways directly involving calcium mobilization – stimulation of PRL or thyroid-stimulating hormone (TSH) secretion by thyrotropin-releasing hormone (TRH) or of all hormonal secretions by phorbol esters. All the effects of SS on pituitary secretions were blocked by pertussis toxin, which inactivates a class of G-proteins including (but not restricted to) the G_i proteins involved in the inhibitory coupling to adenylate cyclase. Thus it was postulated that SS receptors were able to interact not only with G_i proteins but also with G-proteins involved in other mechanisms of action. Yajima et al. [117] demonstrated that phospholipid hydrolysis did not mediate the effects of SS on pituitary cells. Conversely, it was shown that SS in nanomolar concentrations could also reduce the calcium influx in pituitary cells [40, 42] and hyperpolarize their plasma membrane either directly through voltage-dependent calcium channels [108] or indirectly through an increase in potassium conductances [40, 42].

The hyperpolarizing actions of SS, also recorded on brain neurons in the CA1 region of the hippocampus [58] and the solitary tract nucleus [36], were attributed to the opening of a noninactivating, voltage-dependent outward potassium current known as the M current (I_M). Interestingly, this current is inversely regulated by muscarinic cholinergic agonists [22, 36]. Very recently, it was found that the I_M-augmenting effect of SS was blocked by inhibitors of phospholipase A_2 and that arachidonic acid and leukotriene

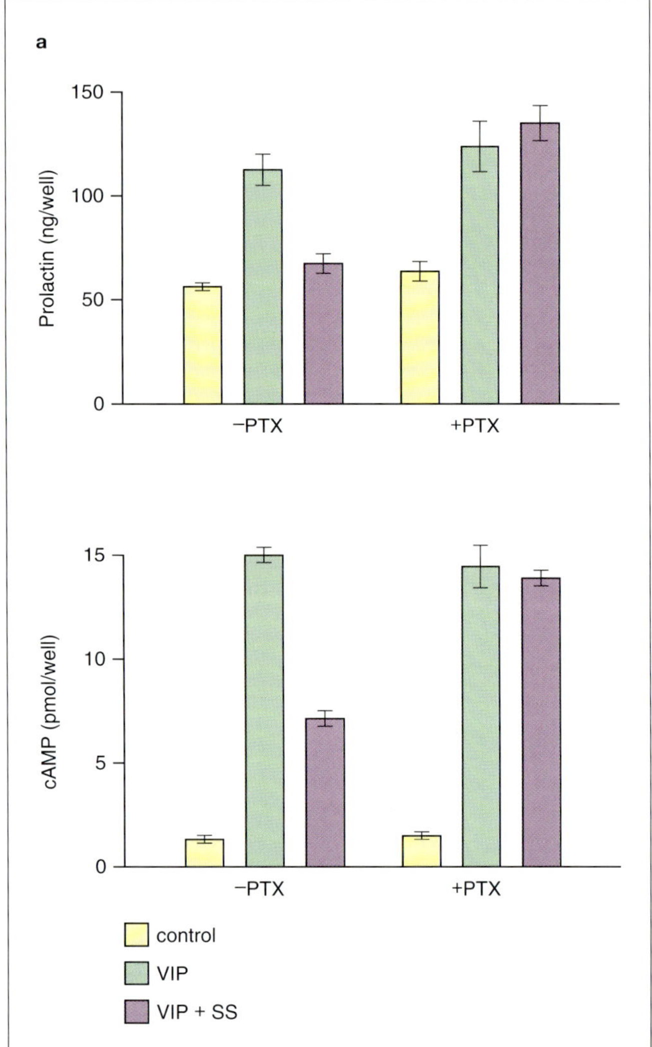

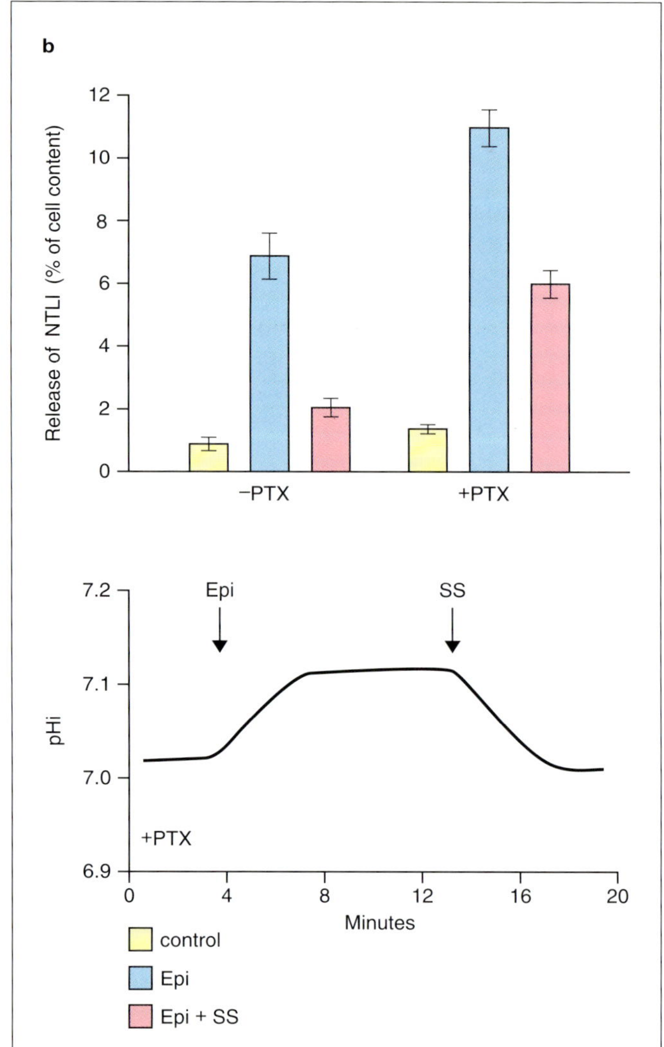

Fig. 5 a, b. *Dual mechanisms of action of SS as revealed by pretreatment with pertussis toxin (**PTX**).* ***a*** *Inhibitory pertussis-toxin-sensitive effects in GH_4C_1 pituitary cells:* ***bottom,*** *cAMP formation;* ***top,*** *PRL secretion. Cells were stimulated by VIP 100 nmol/l and inhibited by SS 100 nmol/l after 24 h of preincubation with or without pertussis toxin 70 ng/ml. Data are mean ± SE of six determinations (modified from [41]).* ***b*** *Inhibitory pertussis-toxin-independent effects in canine enteric endocrine cells:* ***bottom,*** *pH lowering;* ***top,*** *neurotensin-like (**NTLI**) release. Cells were stimulated by 1 µmol/l epinephrine (**Epi**) and inhibited by SS 1 µmol/l after 6 h of preincubation with or without pertussis toxin 300 ng/ml. Data are mean ±SE of seven determinations. (Modified from [2])*

C_4 mimicked the effects of SS on hippocampal pyramidal neurons in vitro [95]. In cardiac [49] and pancreatic [13] cells, other potassium channels have also been implicated in the effects of SS.

In all these studies, SS-28 and SS-14 were equally active. Only in one instance [113] were the two peptides found to elicit opposite effects (hyperpolarization by SS-14 and depolarization by SS-28) on potassium currents on individual cortical neurons in culture. Another study recently reported a depolarizing effect of SS-14 in hippocampal slices through stimulation of L-type calcium channels [57]. The reasons for these discrepancies are still unknown, and it is not yet clear whether they can be related to the pharmacological subtypes SSA and SSB.

SS-induced intracellular reduction in cAMP and calcium concentrations both result in an indirect decrease in protein phosphorylation through their respective specific kinases. Moreover, some of the receptor-mediated effects of SS could be ascribed to a direct inhibition of synaptic plasma membrane protein phosphorylation [14]. Recently, a new class of receptor has been shown to possess a tyrosine phosphatase activity (for review see [34]); if SS receptors belonged to this class, the effects of SS should be dephosphorylation of membrane proteins. Until now, such an effect of SS has been reported only on pancreatic cancer cell preparations [51]. Surprisingly, depending on the cell line, octreotide was active [112] or inactive [51, 91] on the inhibition of cell proliferation. In contrast to its effects on cAMP or on calcium entry, the inhibition by SS of cell proliferation is pertussis-toxin-independent [107], as is the lowering of intracellular pH caused by the peptide in canine enteric endocrine cells [2]. This suggests that the SS receptors participating in these effects are not coupled to G-protein, and this is the strongest argument to date for the existence of different SS receptor subtypes coupled to different mechanisms of action (Fig. 5).

Regulation by Hormones and Somatostatin

Peripheral hormones exert potent influences on specific brain and pituitary cell functions. Part of these regulations involves SS and its receptors.

It has recently been shown that oestradiol positively regulates the synthesis of SS from hypothalamic but not cortical neurons [114]. In the anterior pituitary, the inhibitory effect of SS on PRL (but not GH or TSH) secretion is also dependent on the steroid [110], and the number of SS receptors coupled to the inhibition of adenylate cyclase activity on pituitary lactotrophs is also increased in the presence of oestradiol [38]. A recent report demonstrated an increase in the size of cross-linked pituitary SS receptors in oestradiol-treated membranes [38].

In GH_4C_1 cells, glucocorticoids [92] decrease and thyroid hormones [32] increase the number of SS receptors, and the effects of SS on GH secretion are increased as well. The negative effects of glucocorticoids also affect the inhibition by SS of adenocorticotropic hormone (ACTH) release from normal rat pituitary cells [45]. In vivo, adrenalectomy decreases the SS receptor concentrations in the hypothalamus, striatum and hippocampus, but not in the cortex [86]. Thyroidectomy decreases the SS receptor number in the pituitary and causes GH-secreting cells to dedifferentiate [52].

Changes in the concentrations of endogenous ligands are known to induce modifications in the metabolism of their respective receptors (upregulation or desensitization). Pretreatment with cysteamine, which induces an SS depletion, increases the number and affinity of SS binding sites in cerebrocortical synaptosomes but not in the pituitary [101]. The effect of cysteamine is greater in the hypothalamus than in the hippocampus [5]. Thus, it seems that brain SS receptors can be upregulated in the absence of the endogenous peptide. In contrast, downregulation of brain SS receptors has not been reported in conditions in which SS release is enhanced [31]. In pituitary cell culture, one study reported a decrease in the capacity of SS to inhibit GH and TSH secretion after long-term preincubation in the presence of the peptide [97], but occupancy of the receptor by SS was not ruled out. In vivo, in fasting rats, in which the somatostatinergic fluxes to the anterior pituitary are greatly enhanced, there is no modification of SS receptors [33]. In clonal cell lines, SS pretreatment increased the number of SS binding sites in GH_4C_1 cells and caused no reduction in the responsiveness to SS [68]. In contrast, AtT-20 cells became desensitized to SS [72]. Interestingly, no desensitization to octreotide seems to occur in acromegalic patients treated with the analogue [66].

Multiple Effects

In the pituitary, SS receptors mediate the neurohumoral actions of the peptide on the physiological secretion of GH and TSH; depending on the hormonal conditions, it can also inhibit the secretion of PRL and ACTH.

In the central nervous system, SS has numerous effects on the release and metabolism of other neuropeptides and neurotransmitters (reviews in Table 2 and [17]). These actions are stimulatory or inhibitory, but in most instances a direct relation between the presence of SS receptors and the effects of SS has not been demonstrated as yet. Nevertheless, SS terminals have been visualized on SS dendrites in the periventricular nucleus of the hypothalamus, and anterior hypothalamic SS release can be inhibited presynaptically in autoregulatory fashion. Besides, pretreatment with cysteamine enabled the demonstration to be made that the GHRF mRNA concentration is negatively regulated by SS in a subset of arcuate nucleus neurons bearing SS receptors [5].

The ontogenic development of SS and its receptors could be involved in the maturation of the brain [96]. Indeed, in the cerebellum SS levels [35] and functional receptors [24] are transiently expressed in the first three postnatal weeks; radioautography shows the receptors to be restricted to the external granular cell layer, which corresponds to the matrix in which the stem cells of the granular cerebellar cells undergo proliferation [23]. After the age of 3 weeks, the external granular cell layer disappears concomitantly with SS receptors. A similar transitory expression of SS binding sites is also observed in premigratory areas, such as the germinal epithelium bordering the lateral ventricle, that give rise to the striatum or to the cortical intermediate zone from which the cerebral cortex originates. These observations, together with neurotropic actions on molluscan cells [7, 25] suggest that, in addition to its well-known actions as a neurotransmitter, SS may exert trophic activities involved in brain plasticity.

Like its biochemical effects, the behavioural actions of SS are numerous and contradictory (reviews in Table 3 and [71]). For instance, in the cat and rabbit spinal chord SS is involved in thermal nociception, but analgesia has also been reported in rats and humans. However, most of these studies have not taken into account the opiate-like properties of SS and its agonists. Other behavioural effects (motoricity and cognition) could be related to some physiopathological involvement of SS receptors.

Somatostatin Receptors in Pathology

SS receptors are present in variable amounts in every GH-secreting tumour tested [61, 77], a finding that culminated in the use of SS analogues to treat acromegaly. The number of binding sites has been reported to be lower in tumours from acromegalic patients responding poorly to octreotide therapy [75, 78]. In accordance with the physiological effects of

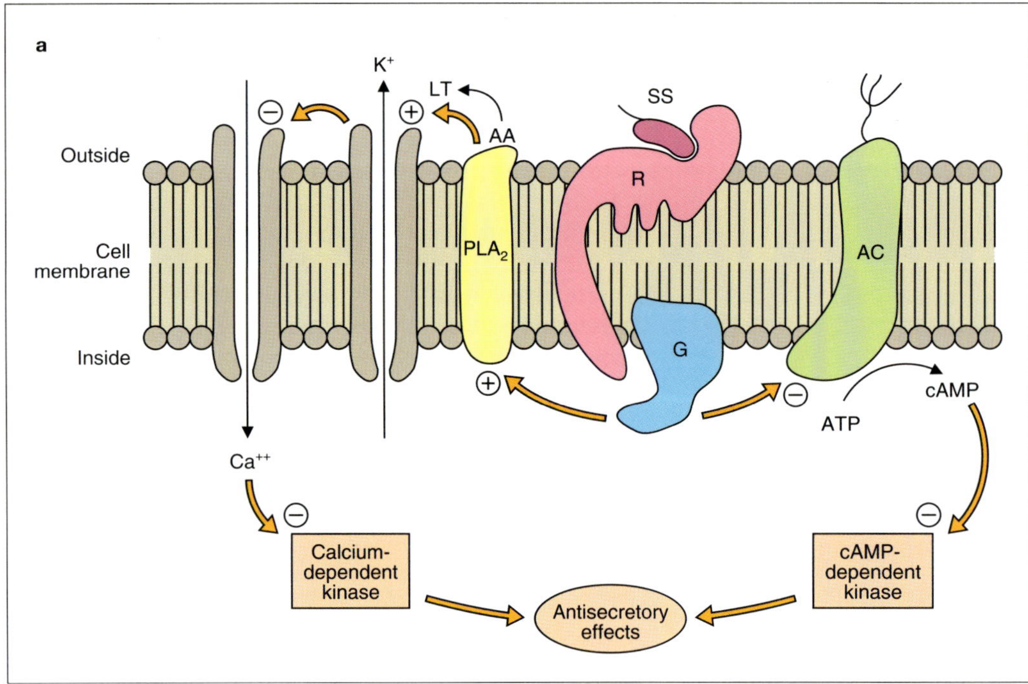

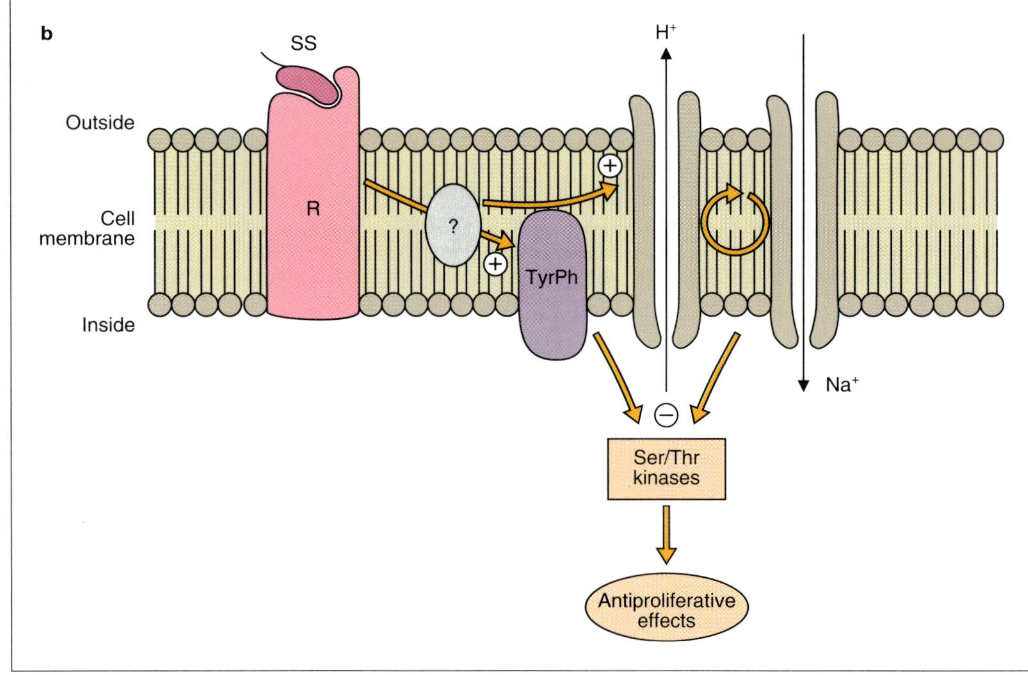

Fig. 6 a, b. *Hypothetical representation of the SS receptors and their multiple signal-transduction pathways.* ***a*** *SS interaction with its receptor (**R**) triggers the inhibition of adenylate cyclase activity (**AC**) and the activation of a potassium conductance (**K^+**) via pertussis-toxin-sensitive guanine-nucleotide-binding protein (**G**). The activation of the K^+ conductance is mediated by the stimulation of phospolipase A_2 (**PLA_2**) through the metabolism of arachidonic acid (**AA**) and the formation of leukotriene (**LT**). The opening of K^+ channels also results in the blocking of calcium (**Ca^{2+}**) entry through voltage-associated Ca^{2+} channels. The inhibition of intracellular Ca^{2+} concentration increase and cyclic adenosine monophosphate (**cAMP**) formation results in the inhibition of their respective kinases. Such a cascade of events is involved in most antisecretory actions of SS.* ***b*** *SS interaction with its receptor (**R**) triggers, via still unknown mechanisms, a tyrosine phosphatase (**TyrPh**) and activates the sodium/proton (**NA/H**) antiporter by reducing the intracellular pH. Both actions could result in the dephosphorylation of Ser/Thr protein kinase substrates involved in early mitogenic events*

the peptide, SS receptors are also present in TSH-secreting tumours and prolactinomas [60]. They are also present in a subset of nonsecreting pituitary tumours. Inhibition of secretion and tumour shrinkage by octreotide have been reported in TSH-secreting adenomas [75] but not in prolactinomas.

SS receptors are located in considerable numbers in meningiomas and glia-derived tumours [76]; interestingly, normal leptomeninges [81] and glia cells [8] present SS binding sites, a fact that could be related to a possible trophic action of SS. However, octreotide has no effect on the basal growth of these tumours or on that stimulated by epidermal growth factor (EGF) [76].

SS receptor alterations have been described in neurodegenerative diseases in which the SS concentrations are modified. In Huntington's disease, striatal SS interneurons are last to degenerate [3], but SS receptors are decreased in the striatum [64]. In Alzheimer's disease, extensive reduction of cortical SS concentrations occurs [12, 87]. A concomitant receptor decrease was initially reported [4] but not confirmed [69], and the current view is that there is a general preservation of SS receptors coupled to inhibition of adenylate cyclase [29]. In view of this preservation, the use of SS-derived drugs in Alzheimer's disease might be expected to afford therapeutic benefits.

Table 2. *Neurochemical effects of SS in the brain*

Brain stucture	*Neurotransmitter or neuropeptide*	*Effect*
Cortex	Acetylcholine	0
	Noradrenaline	+
	Serotonin	0
	Dopamine	+
	Histamine	–
Hippocampus	Acetylcholine	+
	Histamine	–
Striatum	Acetylcholine	0
	Serotonin	0
	Dopamine	+
	Glutamate	–
Hypothalamus	Noradrenaline	–
	Histamine	–
	GHRF	–
	Somatostatin	–
	TRH	–
	CRF	–
Brain Stem	Acetylcholine	0
	Serotonin	0
	Dopamine	+

GHRF, growth-hormone-releasing factor; CRF, corticotropin-releasing factor; TRH, thyrotropin-releasing hormone

Table 3. *Behavioural effects of SS in the brain*

Behaviour	*Injection site*	*Effects*
Body temperature	Intracerebro-ventricular	Increased, unchanged, or decreased
Hunger and satiety	Intracerebro-ventricular	Increased food intake; antagonism of CRF-induced anorexia
	Intrahypothalamic	Decreased food consumption
Nociception	Intracerebro-ventricular	Active on nociception only at high doses that are toxic without a security margin
Sleep	Nucleus of the solitary tract	Increased paradoxical sleep
Motoricity	Intracerebro-ventricular	*Low doses:* decrease in locomotor activity; increased pentobarbital effect; decrease in strychnine-induced seizures
		High doses: catatonia, paraplegia in extension, general seizure, increased mortality
Learning and memory	Intracerebro-ventricular	Attenuates amnesia; inhibits the extinction of active avoidance response

CRF, corticotropin-releasing factor

Conclusion

During the last decade, considerable progress has been achieved in the knowledge of brain SS receptors. A second generation of iodinated ligands, less prone to enzymatic degradation, has enabled numerous in vitro binding assays to be developed. These ligands also made the light-microscopic visualization of the brain receptors possible, and there is now general agreement on their regional distribution and the multiple biochemical and behavioural effects of SS.

The knowledge thus acquired culminated in the design of simpler SS agonists that are now currently used in the clinical management of pituitary adenomas and gastroenteropancreatic (GEP) tumours.

However, much remains to be done. The pharmacology of the SS receptors remains largely unknown owing to the absence of a proper antagonist. The distinction between high- and low-affinity octreotide-sensitive SS binding sites in the brain remains to be correlated with differing transduction mechanisms before these two subtypes can be definitely attributed to different receptors.

The purification of the brain SS receptor is a very competitive domain of research, and the use of molecular biology techniques will certainly result in its structural characterization (see Fig. 6); in turn, this should provide an answer to the question whether the receptors involved in the neurosecretory and the antiproliferative effects of SS are the same or different molecules.

Note added in proof. During the revision of this manuscript, Yamada et al. (Yamada Y, Post SR, Wang K, Tager HW, Bell GI and Seino S [1992] Proc Nat Acad Sci USA 89: 251–255) reported the cloning and sequencing of genes encoding two SS receptors of 391 and 369 amino acids that are members of the seven transmembrane domains superfamily of G-protein-coupled receptors. The two receptors have distinct tissue distributions, the longer one being expressed at higher levels in the jejunum and stomach and the shorter one in the cerebrum and kidney. Both display equivalent affinities for SS-14 and SS-28. This study will give further impetus to the understanding of the molecular basis of the pharmacology and mechanisms of action of SS.

References

1. Aguilera G, Parker DS (1982) Pituitary somatostatin receptors: characterization with a non degradable peptide analogue. J Biol Chem 257: 1134–1137
2. Barber DL, McGuire ME, Ganz MB (1989) β Adrenergic and somatostatin receptors regulate Na/H exchange independent of cAMP. J Biol Chem 264: 21038–21042
3. Beal MF, Benoit R, Bird ED, Martin JB (1985) Immunoreactive somatostatin 28 1–12 is increased in Huntington's disease. Neurosci Lett 56: 377–380
4. Beal MF, Mazurek MF, Tran V, Chattha G, Bird ED, Martin JB (1985) Reduced numbers of somatostatin receptors in the cerebral cortex in Alzheimer's disease. Science 229: 289–291.

5. Bertherat J, Berod A, Normand E, Bloch B, Rostene W, Kordon C, Epelbaum J (1990) Cysteamine induced somatostatin depletion increases SRIF binding and GRF mRNA in the arcuate nucleus of the rat. Neuroendocrinology [Suppl 1] 52: 119
6. Bruno JF, Berelowitz M (1989) Covalent labeling of the somatostatin receptor in the rat anterior pituitary membranes. Endocrinology 124: 831–837
7. Bulloch AGM (1987) Somatostatin enhances neurite outgrowth and electrical coupling of regenerating neurons in helisoma. Brain Res 412: 6–17
8. Chneiweiss H, Bertrand P, Epelbaum J, Kordon C, Glowinski J, Premont J, Enjalbert A (1987) Somatostatin receptors on cortical neurones and adenohypophysis: comparisons between specific binding and adenylate cyclase inhibition. Eur J Pharmacol 138: 249–255
9. Chneiweiss H, Glowinski J, Premont J (1985) Modulation by monoamines of somatostatin-sensitive adenylate cyclase on neuronal and glial cells fromm the mouse brain in primary cultures. J Neurochem 44: 1825–1831
10. Cunningham ET Jr, Sawchenko PE (1989) A circumscribed projection from the nucleus of the solitary tract to the nucleus ambiguus in the rat: anatomical evidence for somatostatin 28 immunoreactive interneurons subserving reflex control of oesophageal motility. J Neurosci 9: 1668–1682
11. Czernik A, Petrack B (1983) Somatostatin receptor binding in rat cerebral cortex, characterization using a non reducible somatostatin analog. J Biol Chem 256: 5525–5530
12. Davies P, Katzmann R, Terry RD (1980) Reduced somatostatin-like immunoreactivity in cerebral cortex from cases of Alzheimer's disease and Alzheimer senile dementia. Nature 288: 279–280
13. De Weille J, Schmid-Antomarchi H, Fosset M, Lazdunski M (1989) Regulation of ATP-sensitive K^+ channels in insulinoma cells: activation by somatostatin and protein kinase C and the role of cAMP. Proc Natl Acad Sci USA 86: 2971–2975
14. Dokas L, Zwiers H, Coy D, Gispen WH (1983) Somatostatin and analogs inhibit endogenous synaptic plasma membrane phosphorylation in vitro. Eur J Pharmacol 88: 185–193
15. Enjalbert A, Rasolonjanahary R, Moyse E, Kordon C, Epelbaum J (1983) Guanine nucleotide sensitivity of 125l-iodo NTyr somatostatin binding in rat adenohypophysis and cerebral cortex. Endocrinology 113: 822–824
16. Enjalbert A, Tapia-Arancibia L, Rieurort M, Moyse E, Kordon C, Epelbaum J (1982) Somatostatin receptors on rat anterior pituitary membranes. Endocrinology 110: 1634–1640
17. Epelbaum J (1986) Somatostatin in the central nervous system: Physiology and pathological modifications. Prog Neurobiol 27: 63–100
18. Epelbaum J (1989) Combined radioautographic and immunohistochemical evidence for an association of somatostatin binding sites with GRF-containing nerve cell bodies in the rat arcuate nucleus J Neuroendocrinology 1: 109–115
19. Epelbaum J, Arancibia LT, Kordon C, Enjalbert A (1982) Characterization, regional distribution, and subcellular distribution of ^{125}l-Tyr1-somatostatin binding sites in rat brain. J Neurochem 38: 1515–1522
20. Epelbaum J, Enjalbert A, Dussaillant M, Kordon C, Rostene WH (1985) Autoradiographic localization of a non reducible somatostatin analog (125l-CGP23996) binding sites in the rat brain: comparison with membrane binding. Peptides 6: 713–719
21. Epelbaum J, Enjalbert A, Krantic S, Musset F, Bertrand P, Rasolon Janahary R, Shu C, Kordon C (1987) Somatostatin receptors on pituitary somatotrophs, thyreotrophs and lactotrophs: pharmacological evidence for loose coupling to adenylate cyclase. Endocrinology 121: 2177–2185
22. Eva C, Costa E (1987) In rat hippocampus, somatostatin 14 and muscarinic receptor ligands modulate an adenylate cyclase belonging to a common domain of the receptor. J Pharmacol Exp Ther 242: 888–894
23. Gonzales BJ, Leroux P, Bodenant C, Laquerriere A, Coy DH, Vaudry H (1989) Ontogeny of somatostatin receptors in the rat brain: biochemical and autoradiographic study. Neuroscience 29: 629–644
24. Gonzales BJ, Leroux P, Laquerriere A, Coy DH, Bodenant C, Vaudry H (1988) Transient expression of somatostatin receptors in the rat cerebellum during development. Dev Brain Res 40: 154–157
25. Grimm-Jorgensen Y (1987) Somatostatin and calcitonin stimulate neurite regeneration of molluscan neurons in vitro. Brain Res 403: 121–126
26. Gulya K, Wamsley JK, Gehlert D, Pelton JT, Duckles SP, Hruby VJ, Yamamura HI (1985) Light microscopic autoradiographic localization of somatostatin receptors in rat brain. J Pharmacol Exp Ther 235: 254–258
27. He HT, Johnson K, Thermos K, Reisine T (1989) Purification of a putative brain somatostatin receptor. Proc Natl Acad Sci USA 86: 1480–1484
28. He HT, Reins-Domiano S, Martin JM, Law SF, Borislow S, Woolkalis M, Manning D, Reisine T (1990) Solubilization of active somatostatin receptors from rat brain. Mol Pharmacol 37: 614–621
29. Heidet V, Slama A, Cervera P, Agid Y, Kordon C, Epelbaum J (1989) Visualization of 125l-Tyr0-D Trp8-SRIF binding sites by quantitative radioautography and immunocytochemichal labelled SRIF containing cells on adjacent sections of human hippocampus and temporal cortex: comparison between control and Alzheimer brain. International Symposium on Somatostatin Aug 6–9, Montreal
30. Heisler S, Reisine T, Hook VYH, Axelrod J (1982) Somatostatin inhibits multireceptor stimulation of cyclic AMP formation and corticotropin secretion in mouse pituitary tumor cells. Proc Natl Acad Sci USA 79: 6502–6506
31. Higuchi T, Kokubu T, Sikand GS, Wada JA, Friesen HG (1984) A study of somatostatin receptors in amygdaloid-kindled rat brain. J Neurochem 43: 1271–1276
32. Hinkle P, Perrone M, Schonbrunn A (1981) Mechanism of thyroid hormone inhibition of thyrotropin-releasing-hormone action. Endocrinology 110: 199–204
33. Hugues JN, Enjalbert A, Moyse E, Shu C, Voirol MJ, Sebaoun J, Epelbaum J (1986) Differential effects of passive immunization with somatostatin antiserum on adenohypophyseal hormone secretions in starved rats J Endocrinol 109: 169–174
34. Hunter T (1989) Protein-tyrosine phosphatases: the other side of the coin. Cell 58: 1013–1016
35. Inagaki S, Shiosaka S, Takatsuki K, Iida H, Sakanaka M, Senba E, Hara Y, Kawa Y, Tohyama M (1982) Ontogeny of somatostatin-containing neurons of the rat cerebellum including its fibers connections: an experimental and immunohistochemical analysis. Dev Brain Res 3: 509–527
36. Jacquin T, Champagnat J, Madamba S, Denavit-Saubie M, Siggins GR (1988) Somatostatin depresses excitability in neurons of the solitary tract complex through hyperpolarization and augmentation of I_M, a non-inactivating voltage dependent outward current blocked by muscarinic agonists. Proc Natl Acad Sci USA 85: 948–952
37. Kimura N (1989) Developmental changes and molecular properties of somatostatin receptors in the rat cerebral cortex. Biochem Biophys Res Commun 160: 72–78
38. Kimura N, Hayafuji C, Kimura N (1989) Characterization of 17β-estradiol dependent and independent somatostatin receptor subtypes in rat anterior pituitary. J Biol Chem 264: 7033–7040
39. Knuhtsen S, Esteve JP, Cambillau C, Colas B, Susini C, Vaysse N (1990) Solubilization and characterization of active somatostatin receptors from rat pancreas. J Biol Chem 265: 1129–1133
40. Koch BD, Blalock JB, Schonbrunn A (1988) Characterization of the cyclicAMP-independent actions of somatostatin in GH cells. I. An increase in potassium conductance in responsible for both the hyperpolarization and the decrease in intracellular free calcium produced by somatostatin. J Biol Chem 263: 216–225
41. Koch BD, Dorflinger L, Schonbrunn A (1985) Pertussis toxin blocks both cyclic AMP-mediated and cyclic AM/P-independant actions of somatostatin. J Biol Chem 260: 13138–13145
42. Koch BD, Schonbrunn A (1988) Characterization of the cyclic AMP-independent actions of somatostatin in GH cells. II. An increase in potassium conductance initiates somatostatin induced inhibition of prolactin secretion. J Biol Chem 263: 226–234
43. Kramer TH, Shook JE, Kazmierski W, Ayres EA, Wire WS, Hru-

by VJ, Burks TF (1989) Novel peptidic Mu opioïd antagonists: Pharmacological characterization in vitro and in vivo. J Pharmacol Exp Ther 249: 544–551

44. Kuhar MJ, Yamamura HI (1975) Light autoradiographic localization of cholinergic muscarinic sites in rat brain by specific binding of a potent antagonist. Nature 253: 560–561
45. Lamberts SW, Zuyderwijk Holder FD, Koetsveld PV, Hofland L (1989) Studies on the conditions determining the inhibitory effect of somatostatin on adrenocorticotropin, prolactin and thyrotropin release by cultured pituitary cells. Neuroendocrinology 50: 44–50
46. Leroux P, Gonzales BJ, Laquerriere A, Bodenant MC, Vaudry H (1988) Autoradiographic study of somatostatin receptors in the rat hypothalamus: validation of GTP-induced desaturatin procedure. Neuroendocrinology 57: 533–544
47. Leroux P, Pelletier G (1984) Radioautographic localization of somatostatin-14 and somatostatin-28 binding sites in the rat brain. Peptides 5: 503–506
48. Leroux P, Quirion R, Pelletier G (1985) Localization and characterization of brain somatostatin receptors as studied with somatostatin-14 and somatostatin-28 receptor radioautography. Brain Res 347: 74–84
49. Lewis DL, Clapham DE (1989) Somatostatin activates an inwardly rectifying K^+channel in neonatal rat atrial cells. Pflugers Arch 41: 492–494
50. Lewis LD, Williams JA (1987) Structural characterization of the somatostatin receptor in rat anterior pituitary membranes. Endocrinology 121: 486–492
51. Liebow C, Reilly C, Serrano M, Schally AV (1989) Somatostatin analogues inhibit growth of pancreatic cancer by stimulating tyrosine phosphatase. Proc Natl Acad Sci USA 86: 2003–2007
52. Markstein R, Stockli KA, Reubi JC (1989) Differential effects of somatostatin on adenylate cyclase as functional correlate for different brain somatostatin receptor subpopulations. Neurosci Lett 104: 13–18
53. Martin D, Epelbaum J, Bluet-Pajot MT, Prelot M, Kordon D, Durand D (1985) Thyroidectomy abolishes pulsatile growth hormone secretion without affecting hypothalamic somatostatin. Neuroendocrinology 41: 476–481
54. Maurer R, Gaehwiler BH, Buescher HH, Hill RC, Roemer D (1982) Opiate antagonistic properties of an octapeptide somatostatin analog. Proc Natl. Acad Sci USA 79: 4815–4817
55. Maurer R, Reubi JC (1985) Brain somatostatin receptor subpopulation visualized by autoradiography. Brain Res 333: 178–181
56. McCarthy R, Plunkett LM (1987) Quantitative autoradiographic analysis of somatostatin binding sites in discrete areas of rat forebrain. Brain Res Bull 18: 29–34
57. Miyoshi R, Kito S, Katayama S, Kim SU (1989) Somatostatin increases intracellular calcium concentration in cultured rat hippocampal neurons. Brain Res 489: 361–364
58. Moore SD, Madamba SG, Joels M, Siggins GR (1988) Somatostatin augments the M-current in hippocampal neurons. Science 239: 278–280
59. Morel G, Leroux P, Pelletier G (1989) Ultrastructural autoradiographical localization of somatostatin-28 in the rat pituitary gland. Endocrinology 116: 1615–1623
60. Moyse E, Benoit R, Enjalbert A, Gautron JP, Kordon C, Ling N, Epelbaum J (1984) Subcellular distribution of somatostatin-14, somatostatin-28 and somatostatin-28 (1–12) in rat brain and comparison of their respective binding sites in brain and pituitary. Regul Pept 9: 129–137
61. Moyse E, Le Dafniet M, Epelbaum J, Pagesy P, Peillon F, Kordon C, Enjalbert A (1985) Somatostatin receptors in human growth hormone and prolactin secreting pituitary adenomas. J Clin Endocrinol Metab 65: 65–73
62. Moyse E, Slama A, Videau C, d' Angela P, Kordon C, Epelbaum J (1989) Regional distribution of somatostatin receptor affinity states in rat brain: effects of divalent cations and GTP. Regul Pept 26: 225–234
63. Murthy KK, Srikant CB, Patel YC (1989) Evidence for multiple protein-constituents of the somatostatin receptor in pituitary tumor cells: Affinity cross-linking and molecular characterization. Endocrinology 125: 948–955
64. Palacios JM, Rigo M, Chiniglia G, Probst A (1990) Reduced density of striatal somatostatin receptors in Huntington's chorea. Brain Res 522: 342–346
65. Patel YC, Baquiran G, Srikant CB, Posner BI (1986) Quantitative in vivo autoradiographic localization of 125I-Tyr11-Somatostatin 14 and 125I-Leu8-DTrp22-Tyr25-somatostatin 28 binding sites in rat brain. Endocrinology 119: 2262–2269
66. Plewe G, Schrezebmeir J, Nolken G, Krause U, Beyer J, Kasper H and del Pozo E (1986) Long term treatment of acromegaly wilth the somatostatin analogue SMS201995 over 6 months. Klin Wochenschr 64: 389–392
67. Presky DH, Schonbrunn A (1986) Receptor bound somatostatin and epidermal growth factor are processed differently in GH4C1 rat pituitary cells. J Cell Biol 102: 878–888
68. Presky DH, Schonbrunn A (1988) Somatostatin pretreatment increases the number of somatostatin receptors in GH4C1 pituitary cells and does not reduce cellular responsiveness to somatostatin. J Biol Chem 263: 714–721
69. Quirion R, Martel JC, Robitaille Y, Etienne P, Wood P, Nair NPV, Gauthier S (1986) Neurotransmitter and receptor deficits in senile dementia of the Alzheimer type. Can J Neurol Sci 13: 503–510
70. Raynor K, Reisine T (1989) Analogs of somatostatin selectively label distinct subtypes of somatostatin receptors in brain. J Pharmacol Exp Ther 251: 510–517
71. Reichlin S (1983) Somatostatin I and II. N Engl J Med 309: 1495–1563
72. Reisine T, Axelrod J (1983) Prolonged somatostatin pretreatment desensitizes somatostatin's inhibition of receptor-mediated release of adrenocorticotropin hormone and sensitizes adenylate cyclase. Endocrinology 113: 811–813
73. Reubi JC (1984) Evidence for two somatostatin receptor types in rat brain cortex. Neurosci Lett 49: 259–263
74. Reubi JC, Cortes R, Maurer R, Probst A, Palacios JM (1986) Distribution of somatostatin receptors in the human brain: an autoradiographic study. Neuroscience 18: 329–346
75. Reubi JC, Heitz PU, Landolt AM (1988) Visualization of somatostatin receptors and correlation with immunoreactive growth hormone and prolactin in human pituitary adenomas. J Clin Endocrinol Metab 65: 65–73
76. Reubi JC, Horisberger V, Lang W, Koper JW, Braakman R, Lamberts SW (1989) Coincidence of EGF receptors and somatostatin receptors in meningiomas but inverse differentiation-dependent relationship in glial tumors. Am J Pathol 134: 337–344
77. Reubi JC, Landolt AM (1984) High density of somatostatin receptors in pituitary tumors from acromegalic patients. J Clin Endocrinol Metab 59: 1148–1151
78. Reubi JC, Landolt AM (1988) The growth hormone response to octreotide in acromegaly correlate with adenoma somatostatin receptor status. J Clin Endocrinol Metab 68: 844–850
79. Reubi JC, Maurer R (1985) Autoradiographic mapping of somatostatin receptors in the rat CNS and pituitary. Neuroscience 15: 1183–1193
80. Reubi JC, Maurer R (1986) Different ionic requirements for somatostatin receptors subpopulations in brain. Peptides 14: 301–311
81. Reubi JC, Maurer R, Lamberts SW (1986) Somatostatin binding sites in human leptomeninx. Neurosci Lett 70: 183–186
82. Reubi JC, Perrin M, Rivier J, Vale W (1982) High affinity binding sites for somatostatin to rat pituitary. Biochem Biophys Res Commun 105: 1538 1545
83. Reubi JC, Probst A, Cortes R, Palacois JM (1987) Distinct topographical localization of two somatostatin receptor subpopulations in the human cortex. Brain Res 406: 391–396
84. Reyl Desmars F, Leroux S, Linard S, Benkouka F, Lewin MJM (1989) Purification jusqu' à apparente homogénéité du récepteur de la somatostatine de la lignée cellulair humaine HGT-1 d'origine gastrique. C R Acad Sci III 308: 251–254
85. Reyl Desmars F, Leroux S, Linard S, Benkouka F, Lwin MJM (1989) Solubilization and immunopurification of a somatostatin receptor from the human gastric tumoral cell line HGT-1. J Biol Chem 264: 18789–18795
86. Rodriguez MN, Gomez-Pan A, Arilla E (1989) Decrease in number of somatostatin receptors in rat brain after adrenalectomy: nor-

malization after glucocorticoid replacement. Endocrinology 123: 1147–1152
87. Rossor M, Emson P, Iversen L, Mountjoy C, Roth M, Fahrenkrug J, Rehfeld J (1982) Neuropeptides and neurotransmitters in cerebral cortex in Alzheimer's disease. In: Corkin S, et al. (eds) Alzheimer's disease: a report of progress. Raven, New York, pp 15–23
88. Rouleau D, Barden N (1981) Inhibition of anterior pituitary prostaglandin stimulated adenylate cyclase activity by somatostatin. Can J Biochem 59: 307–310
89. Sakamoto C, Goldfine ID, Williams JA (1984) The somatostatin receptor on isolated pancreatic acinar cell plasma membranes. J Biol Chem 259: 9623–9627
90. Sakamoto C, Nagao M, Matosaki T, Nishizaki H, Konda Y, Baba S (1988) Somatostatin Receptors on rat cerebrocortical membranes. J Biol Chem 263: 14441–14445
91. Schally AV, Comura Schally AM, Redding T (1984) Antitumor effects of analogs of hypothalamic hormones in endocrine dependent cancers. Proc Soc Exp Biol Med 175: 259–281
92. Schonbrunn AH (1982) Glucocorticoids down-regulate somatostatin receptors on pituitary cells in culture. Endocrinology 110: 1147–1152
93. Schonbrunn AH, Rorstad OP, Westendorf JM, Martin JB (1983) Somatostatin analogs: correlation between receptor binding affinity and biological potency in GH pituitary cells. Endocrinology 113: 1559–1567
94. Schonbrunn AH, Tashjian AJR (1978) Characterization of functional receptors for somatostatin in rat pituitary cells in culture. J Biol Chem 253: 6473–6483
95. Schweizer P, Madamba S, Siggins GR (1990) Arachidonic acid metabolites as mediators of somatostatin-induced increase of neuronal M-current. Nature 346: 464–466
96. Shiosaka S, Takatsuki K, Sakanaka M, Inagaki S, Tkagi H, Senba E, Kawai Y, Iida H, Minagawa M, Mara H, Matsuzaka T, Tohyama M (1982) Ontogeny of somatostatin-containing neuron system of the rat: immunohisto-chemical analysis. II. Forebrain and diencephalon. J Comp Neurol 204: 211–224
97. Smith M, Yamamoto G, Vale W (1984) Somatostatin desensitization in rat anterior pituitary cells. Mol Cell Endocrinol 37: 311–318
98. Srikant CB, Patel YC (1981) Somatostatin receptors. Identification and characterization in rat brain membranes Proc Natl Acad Sci USA 78: 3930–3934
99. Srikant CB, Patel YC (1981) Receptor binding of somatostatin 28 is tissue specific. Nature 294: 259–260
100. Srikant CB, Patel YC (1982) Characterization of pituitary membrane receptors for somatostatin in the rat. Endocrinology 110: 2138–2146
101. Srikant CB, Patel YC (1984) Cysteamine-induced depletion of brain somatostatin is associated with up-regulation of cerebrocortical somatostatin receptors. Endorcrinology 115: 990–995
102. Srikant CB, Patel YC (1986) Chemical cross linking of somatostatin receptors in rat adrenal cortex. Biochem Biophys Res Commun 139: 757–762
103. Srikant CB, Patel YC (1986) Somatostatin receptors on rat pancreatic acinar cells. J Biol Chem 261: 7690–7696
104. Susini C, Bailey A, Szecowka J, Williams JA (1986) Characterization of covalently-cross-linked pancreatic somatostatin receptors. J Biol Chem 261: 16738–16743
105. Terenius L (1976) Somatostatin and ACTH are peptides with partial antagonist-like selectivity for opiate receptors. Eur J Pharmacol 38: 211–213
106. Thermos K, He HT, Wang HL, Margolis N, Reisine T (1989) Biochemical properties of brain somatostatin receptors. Neuroscience 31: 131–141
107. Tran VT, Beal MF, Martin JB (1985) Two types of somatostatin receptors differentiated by cyclic somatostatin analogs. Science 228: 492–495
108. Tsunoo A, Yoshii M, Narahashi T (1986) Block of calcium channels by enkephalin and somatostatin in neuroblastoma-glioma hybrid NG108-15 cells. Proc Natl. Acad Sci USA 83: 9832–9836
109. Uhl GR, Tran V, Snyder SH, Martin JB (1985) Somatostatin receptors: distribution in rat central nervous system and human frontal cortex. J Comp Neurol 240: 288–304
110. Vale W, Rivier C, Brazeau P, Guillemin R (1974) Effects of somatostatin on the secretion of thyrotropin and prolactin. Endocrinology 95: 968–977
111. Van Calker D, Muller M, Hamprecht B (1980) Regulation by secretin, VIP and somatostatin of cyclicAMP accumulation in cultured brain cells. Proc Natl Acad Sci USA 77: 6907–6911
112. Viguerie N, Tahiri-Jouhi N, Ayral AM, Cambillau C, Scemama JL, Bastié JM, Knuhtsen S, Esteve JP, Pradayrol L, Susini C, Vaysse N (1989) Direct inhibitory effects of a somatostatin analog, SMS201995, on ARA42J cell proliferation via a pertussis toxin insensitive guanosine-triphosphate binding protein independent mechanism. Endocrinology 124: 1017–1025
113. Wang HL, Bogen C, Reisine T, Dichter M (1989) Somatostatin 14 and somatostatin 28 induce opposite effects on, potassium currents in rat neocortical neurons. Proc Natl Acad Sci USA 86: 9616–9620
114. Werner H, Koch Y, Baldino F Jr, Gozes I (1988) Steroid regulation of somatostatin mRNA in the hypothalamus. J Biol Chem 263: 7666–7671
115. White M, Reisine T (1990) Expression of functional pituitary somatostatin receptors in xenopus oocytes. Proc Natl Acad Sci USA 87: 133–136
116. Whitford CA, Candy JM, Snell CR, Hirst BH, Oakley AE, Johnson M, Thompson JE (1987) Autoradiographic vizualization of binding sites for (^{3}H)somatostatin in the rat brain. Eur J Phamacol 138:327–333
117. Yajima Y, Akita Y, Saito T (1986) Pertussis toxin blocks the inhibitory effects of somatostatin on cAMP-independent thyrotropin hormone releasing hormone stimulated prolactin secretion of GH3 cells. J Biol Chem 261: 2684–2691

Somatostatin in Neuropsychiatric Disorders

D. R. Rubinow, C. L. Davis, and R. M. Post

National Institute of Mental Health, Biological Psychiatry Branch, Bethesda, MD 20892, USA

Introduction

Early efforts to identify the hypothalamic stimulating factor for growth hormone (GH) release from the pituitary inadvertently resulted in the discovery of a hypothalamic inhibitor of GH release [57]. Despite initial skepticism regarding its existence, this inhibiting factor, somatostatin (SS; also called somatotropin release-inhibiting factor, SRIF), was subsequently isolated and structurally characterized by Brazeau et al. [17]. As has been true for most of the brain neuropeptides identified over the last two decades, the known neuroregulatory effects of the tetradecapeptide SS have expanded far beyond its originally discovered role. The possibility that alterations in SS might reflect or contribute to the pathophysiology of disorders of the central nervous system (CNS) was first examined by Patel et al. in 1977 [75]. Subsequently, a variety of reports appeared describing decreased SS in brain and/or cerebrospinal fluid (CSF) of patients with several neuropsychiatric disorders, particularly Alzheimer's disease and depression. These studies of SS in neuropsychiatric disorders will be reviewed in this chapter, as will the evidence supporting a role for SS as a modulator of CNS activity.

Neuropsychiatric Disease-Related Alterations

In 1977, Patel et al. [75] provided the first report of disease-related alterations in SS levels in the cerebrospinal fluid (CSF). They described increased CSF levels of SS, compared with neurological controls, in several inflammatory or destructive neurological disorders (spinal cord disease, nerve root compression, cerebral tumor, meningitis, metabolic encephalopathy). Similar findings were also reported by Beal et al. [5]. Presumably, the elevated CSF levels of SS observed by these groups resulted from "leakage" from damaged or anoxic neurons.

In contrast, decreased CSF SS levels, perhaps more suggestive of functional neuronal alterations, have been noted in several neuropsychiatric disorders including Parkinson's disease [32], delirium [55], and multiple sclerosis (MS) during relapse [5, 116]. Su et al. [121] recently confirmed decreased SS in progressive MS (in contrast to the findings of Rosler et al. [100]), and further demonstrated a significant decrease in SS in serial CSF samples over 2 years in a small group of medication-free patients with MS who manifested neurological deterioration during the same interval. Koponen et al. [55] observed significantly decreased SS in multi-infarct dementia (MID) patients with delirium, consistent with the decreased levels in MID seen by Beal et al. [4] and Bissette et al. [13]; however, nonsignificant decreases were observed by Cramer et al. [23]. Decreased CSF SS has also been reported recently in patients with adrenocorticotropic-hormone- (ACTH-) dependent Cushing's syndrome [54]. Findings in patients with Huntington's disease [4] have been inconsistent, with both decreased and normal levels observed.

Alzheimer's Disease

Highly consistent findings of low CSF SS have been reported in senile dementia [13, 73] and Alzheimer's disease [4, 23, 27, 38, 42, 108, 113, 114, 122, 140]. The results of human postmortem studies to date parallel those in CSF. Markedly decreased SS concentrations have been demonstrated in a variety of cortical and subcortical sites in Alzheimer's disease [19, 101]. The amount of the reduction and the brain regions most clearly affected differ across studies. Lowe et al. [61] suggest that these discrepancies may reflect differences in the length and severity of disease; thus, studies with a greater representation of severely ill patients [7] report greater decreases in brain SS than are seen in other studies [101, 123]. Further, differences may reflect postmortem changes, molecular heterogeneity of SS-like immunoreactivity, or decreases in SS masked by the loss of non-SS tissue volume [19]. Nonetheless, SS concentrations are widely (but not consistently [19]) reported to be reduced to 21%–67% of control values in the temporal cortex, with similar reductions seen in regions of the frontal and parietal but not occipital or cingulate cortex [33, 60, 101, 123]. Decreased SS concentrations have been reported as well in the cortex and hippocampus in Parkinson's disease [33], while increases have been described in the basal ganglia of patients with Huntington's disease [6].

The decreased brain and CSF SS levels in Alzheimer's disease are now highly replicated findings (Fig. 1) and, with

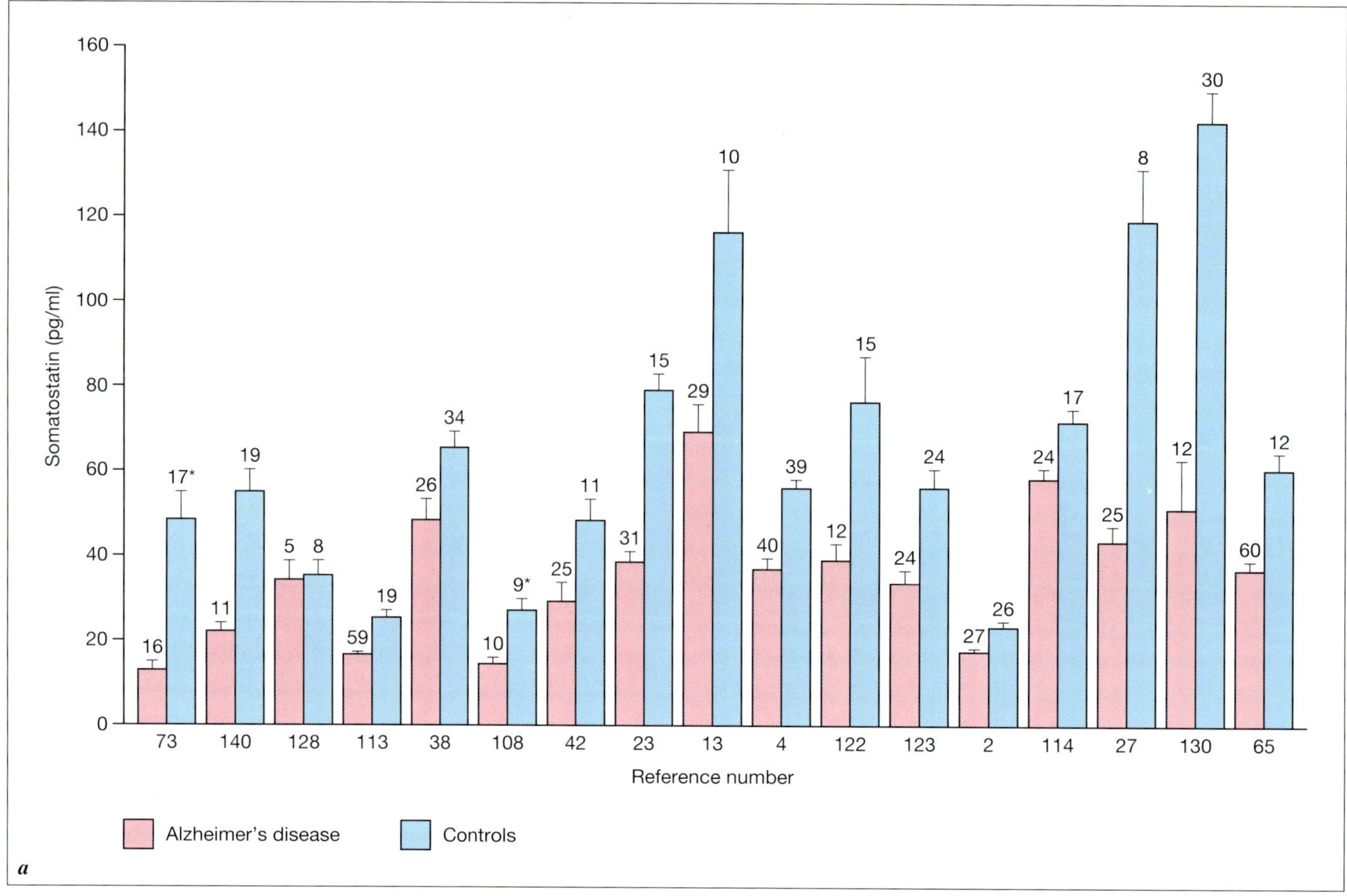

Fig. 1 a, b. *Summary of 17 studies of CSF SS in Alzheimer's disease.*
a *Absolute SS levels in patients and controls. The numbers of subjects and controls appear above the standard error bars. Studies are presented in chronological order. Asterisks indicate neurological rather than normal controls*

evidence of cholinergic neuron degeneration, are among the most commonly observed biochemical abnormalities seen in this disorder. The clinical relevance of altered SS function in Alzheimer's disease is also suggested by the following: (a) reports of correlations between the degree of cognitive impairment and reduction of SS levels in brain [37] and CSF [113]; (b) reports of the colocalization of SS-like immunoreactivity with senile plaques [66] and neurofibrillary tangles [97]; (c) observation of a correlation between the loss of SS and disease severity as determined by plaque density [78]; (d) the demonstration that the brain areas showing the greatest reduction in SS are those that are most markedly hypometabolic as determined by positron emission tomography (PET) scan [123]; and (e) the observation [138] of relative increase in CSF SS concentration in association with a reversal of dementia symptoms in patients with Alzheimer's disease exposed to intensive environmental stimulation. This last finding is of interest given the presumption that the changes in brain SS in Alzheimer's disease reflect neuronal degeneration. It is further noteworthy given the lack of therapeutic success associated with attempts to manipulate pharmacologically central SS levels (by administering a SS analogue to patients with Alzheimer's disease [24] or a SS-depleting agent, i.e., cysteamine, to patients with Huntington's disease [112]). The extent to which these treatments produced changes in central SS levels is uncertain.

Depression

Among the psychiatric disorders studied, the clearest evidence of an alteration in central SS concentration and/or function is in affective illness (Fig. 2). Gerner and Yamada [40] and Rubinow et al. [103] reported significant decreases in CSF SS in medication-free depressed patients compared with normal controls. These initial findings were subsequently replicated. Agren and Lundqvist [1] found significantly lower SS levels in a group of 21 depressed patients studied during their worst week of depression compared with a group of 25 depressed patients studied more than 2 months after their most depressed week. Lower levels of SS were reported by Black et al. [14] in the ventricular CSF of 18 medicated depressed patients (studied prior to cingulotomy) compared with ventricular CSF values in nine pa-

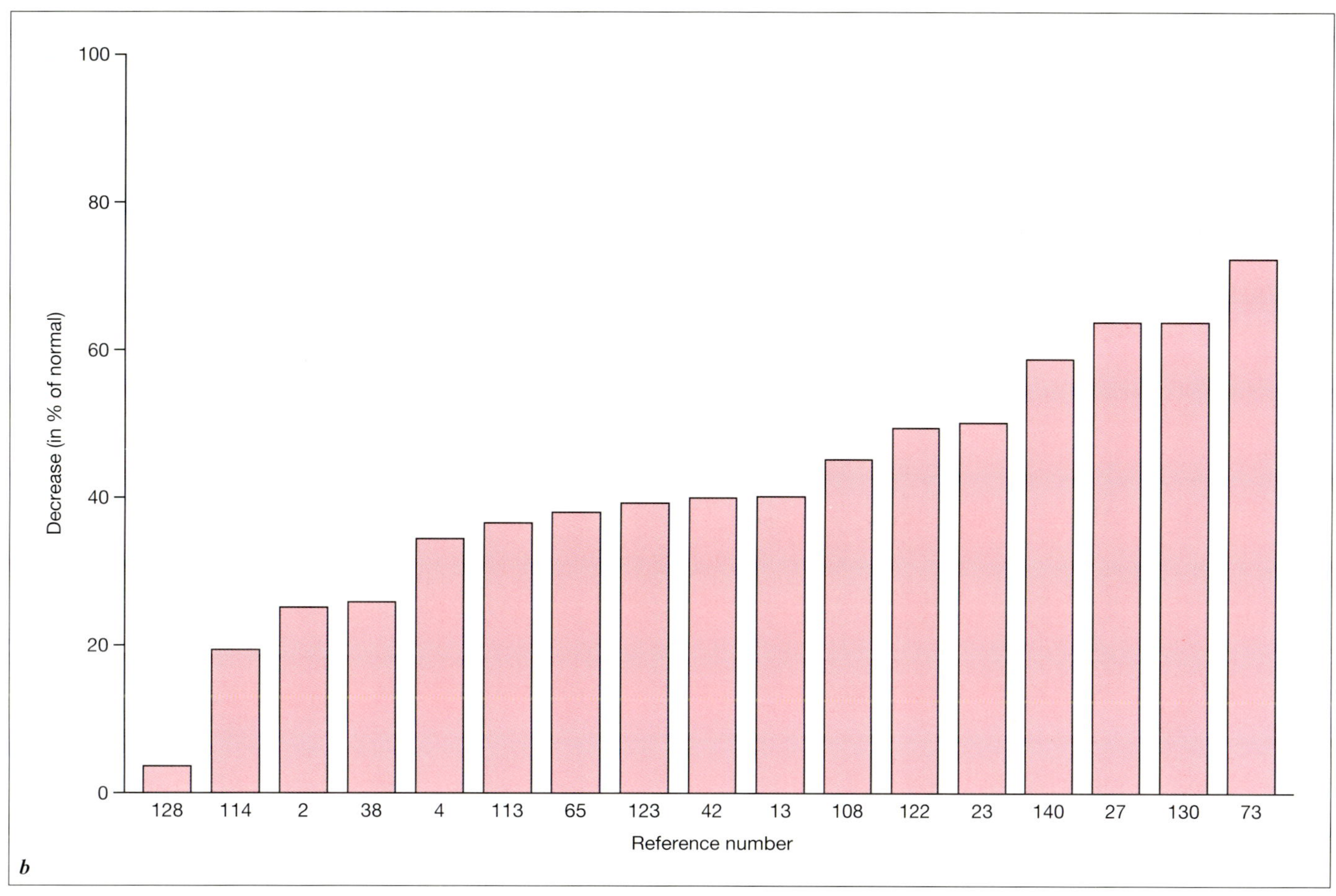

b *The percentage decrease in CSF SS in Alzheimer's disease in the same studies, presented in order of increasing difference between patients and controls.*

tients with hydrocephalus and lumbar CSF values in 46 normal controls. Bissette et al. [13] observed lower CSF SS levels in 17 medication-free depressed patients compared with ten normal controls. Sunderland et al. [122] observed decreased CSF SS in elderly depressed patients and patients with Alzheimer's disease compared with age-matched controls. More recently, in an expanded study, this group described significantly lower CSF SS in 18 elderly depressed patients and in 60 patients with Alzheimer's disease compared with 12 age-matched controls [65]. This group further found a significant negative correlation between the degree of depressed mood in the patients with Alzheimer's disease and CSF SS levels, as well as a trend toward a negative correlation between SS and Hamilton depression scale scores (i.e. lower SS associated with more severe depression ratings) in the patients with major depression. Among patients with Alzheimer's disease, those who were also depressed had significantly lower CSF SS than did nondepressed patients. Davis et al. [27] also observed significantly lower CSF SS levels in eight depressed patients and 25 patients with Alzheimer's disease compared with eight controls, with no significant differences noted between SS levels in the two patient groups. Finally, in an extension of earlier findings, Rubinow [102] reported significantly decreased SS levels in 49 medication-free depressed patients compared with 47 normal volunteers and 23 patients with affective illness in the improved or euthymic state (Fig. 3). Additionally, depression-related decreases in CSF SS were seen in nine medication-free patients who received lumbar punctures during both depressed and improved or manic states; i.e., the values obtained during depression were significantly lower than those obtained during the other state. Thus, decreased CSF SS levels appear to be state related and to normalize with recovery from depression.

In contrast to the consistent findings in studies of depressed patients, investigations of CSF SS in other psychiatric populations have yielded conflicting or as yet unconfirmed results. Both increased [40] and normal [103] SS levels have been reported in manic patients. Bissette et al. [13] observed lower CSF SS concentrations in 11 schizophrenics compared with controls, while Doran et al. [30] found normal levels in 44 schizophrenics. Both decreased [40] and normal SS levels [53] have also been described in patients with anorexia nervosa. In the study by

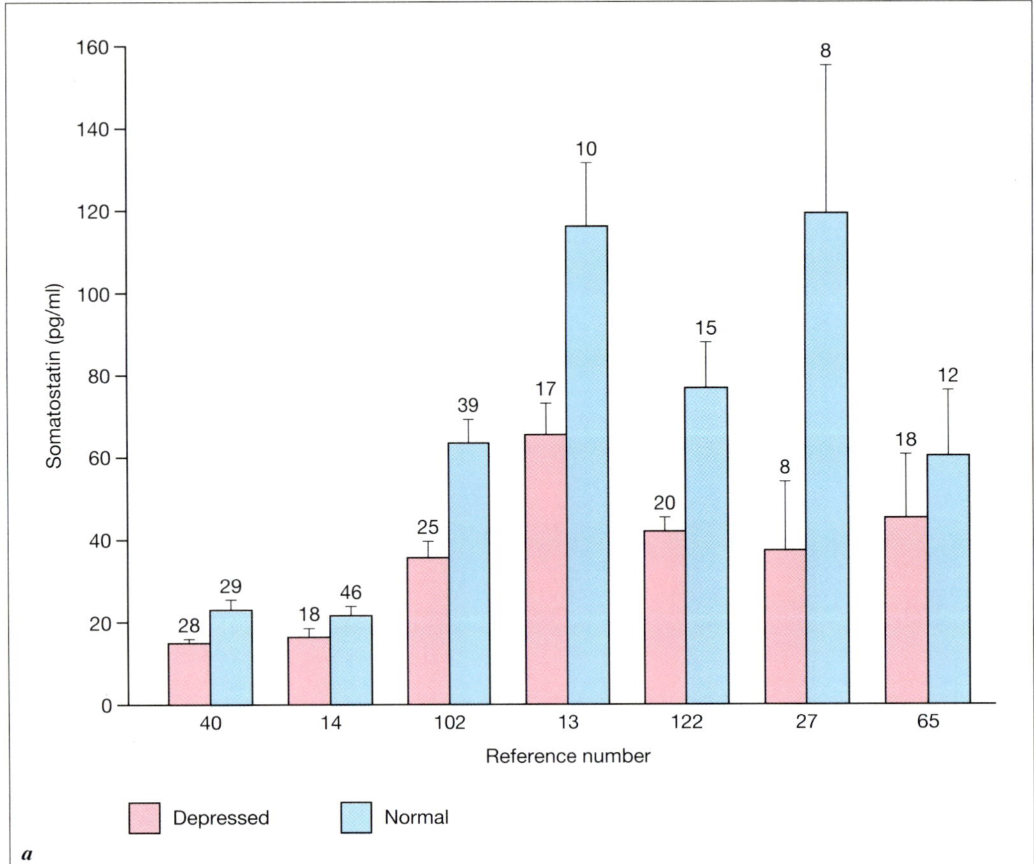

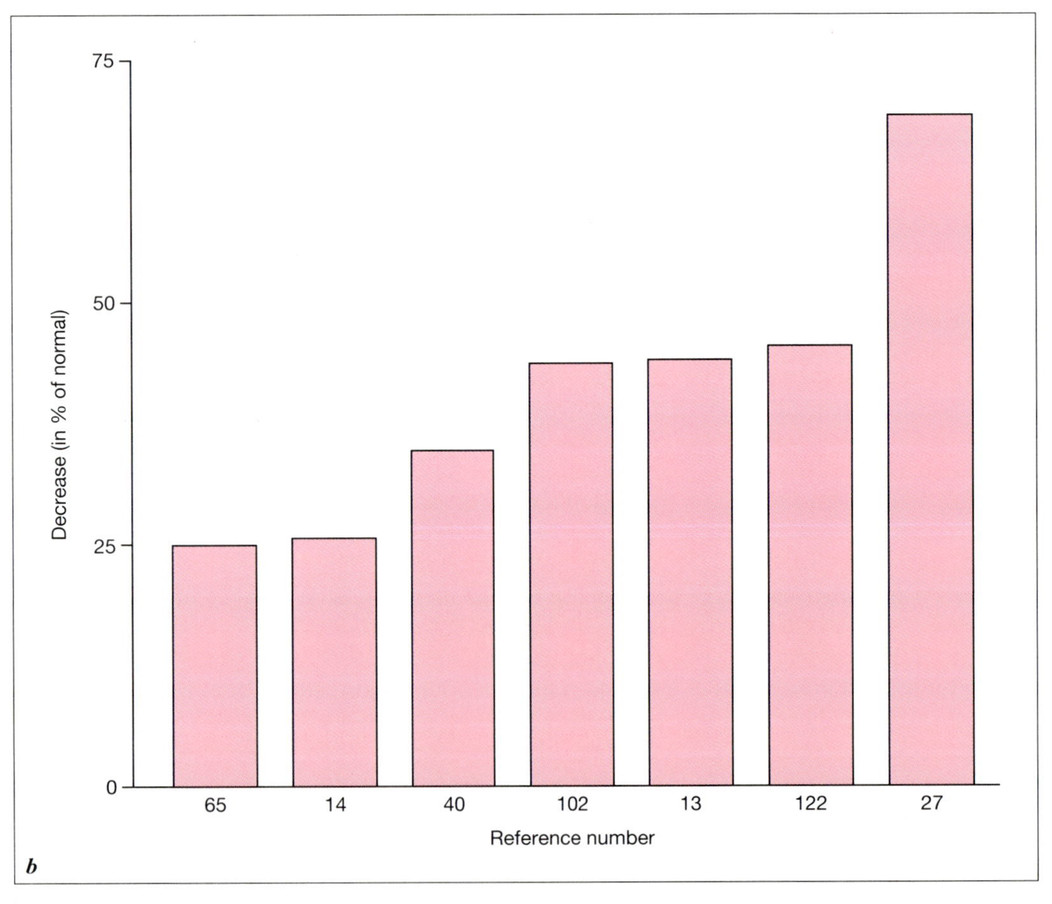

Fig. 2 a, b. *Summary of seven studies showing decreased CSF SS in depression.* ***a*** *Absolute SS levels in patients and controls. The studies are presented in chronological order. The data in [13] have been slightly extended in [137]. The numbers of subjects and controls appear above the standard error bars.* ***b*** *The percentage decrease in CSF SS in depression in the same studies, presented in order of increasing difference between patients and controls.*

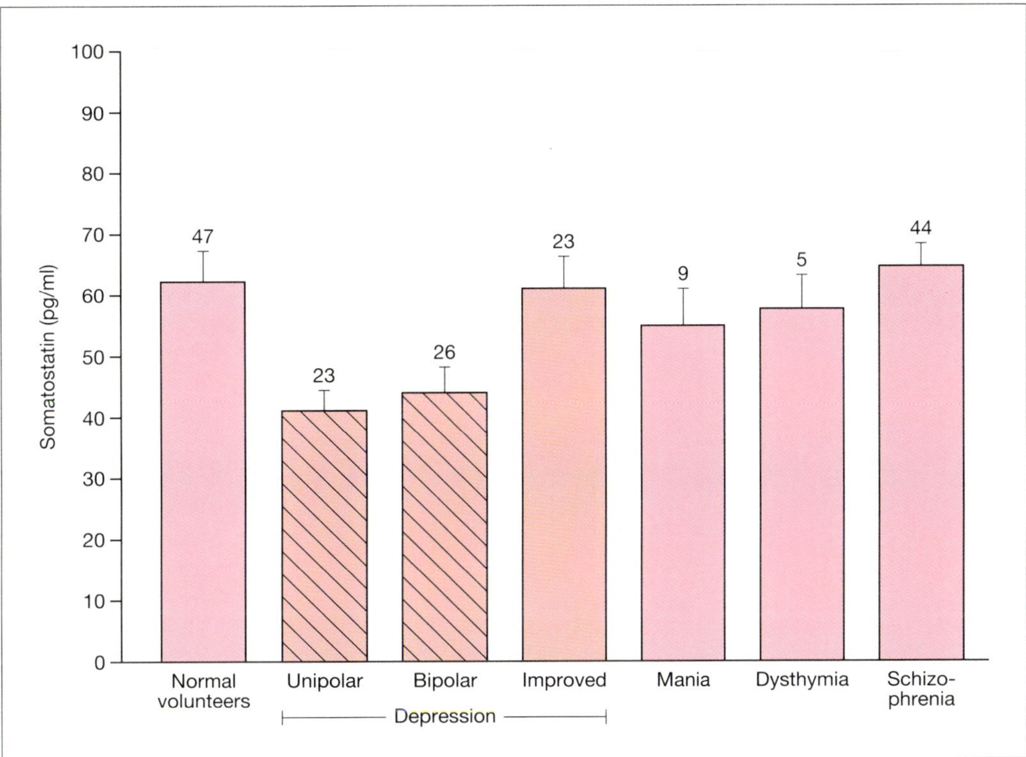

Fig. 3. CSF SS is significantly decreased in depression relative to normal volunteers, affective disorder patients during the improved state, and schizophrenic patients. The numbers of subjects appear above the standard error bars. (Adapted from [102])

Kaye et al. [53], a small but significant increase in CSF SS was noted in patients with bulimia during abstinence from bingeing compared with the same patients when actively bingeing. Normal levels of CSF SS in panic disorders were reported by Widerlov and colleagues, who also observed normal SS levels in the premenstrual syndrome (PMS) as well as no effect of the menstrual cycle phase on SS in patients suffering from PMS and controls [131]. Finally, in the only study of CSF SS in children to date, Kruesi et al. [56] observed significantly lower SS levels in ten patients with disruptive behavior disorder than in ten age- and sex-matched obsessive compulsive children.

Only a few postmortem studies of brain SS in patients with psychiatric illnesses have been performed. In schizophrenics, increases in SS concentration in the lateral thalamus [98] and reductions in the frontal cortex [71] and in the hippocampus (in negative-symptom or type II schizophrenics) have been described [34]. Negative symptom ratings were significantly correlated with hippocampal SS content in this last study. Charlton et al. [20] found no difference in SS concentration or SS receptor affinity and binding capacity in the temporal or occipital cortex in a group of nine primarily medicated depressed patients compared with controls. They similarly found no difference in SS immunoreactivity in the frontal cortex in depressed patients [35] or in the frontal, motor, parietal, or temporal cortex of 12 suicide victims compared with controls. A significant decrease in SS content was observed by Bowen et al. [15] in the temporal pole and pars opercularis of the frontal lobe (but not in other brain regions examined) in seven depressed and previously medicated patients. Finally, no differences in SS content were observed in the frontal cortex, amygdala, or caudate in schizophrenic or affectively ill suicide victims compared with accident victim controls (Davis et al., unpublished data). Clearly, further studies of brain SS in autopsy specimens of depressed patients are required in order to elucidate the potential central concomitants of the decreased CSF SS.

Central Nervous System Effects of Somatostatin

Interpretation of the disease-related alterations in SS is best accomplished in the context of the considerable evidence suggesting a role for SS as a modulator of CNS activity.

Central Nervous System Localization

SS nerve terminals are distributed in a widespread and selective fashion throughout the brain, appearing in the median eminence (the source of anterior pituitary regulation), the posterior pituitary, the limbic system, the cortex, the hippocampus, and the hypothalamus and hypothalamic projections to the brain stem and spinal cord [87]. SS is most concentrated in the hypothalamus, while it appears in the greatest abundance in the cortex. SS-containing cell bodies have been identified in diverse brain regions including the neocortex, amygdala, hippocampus, preoptic and periventricular nuclei, striatum, nucleus accumbens, locus

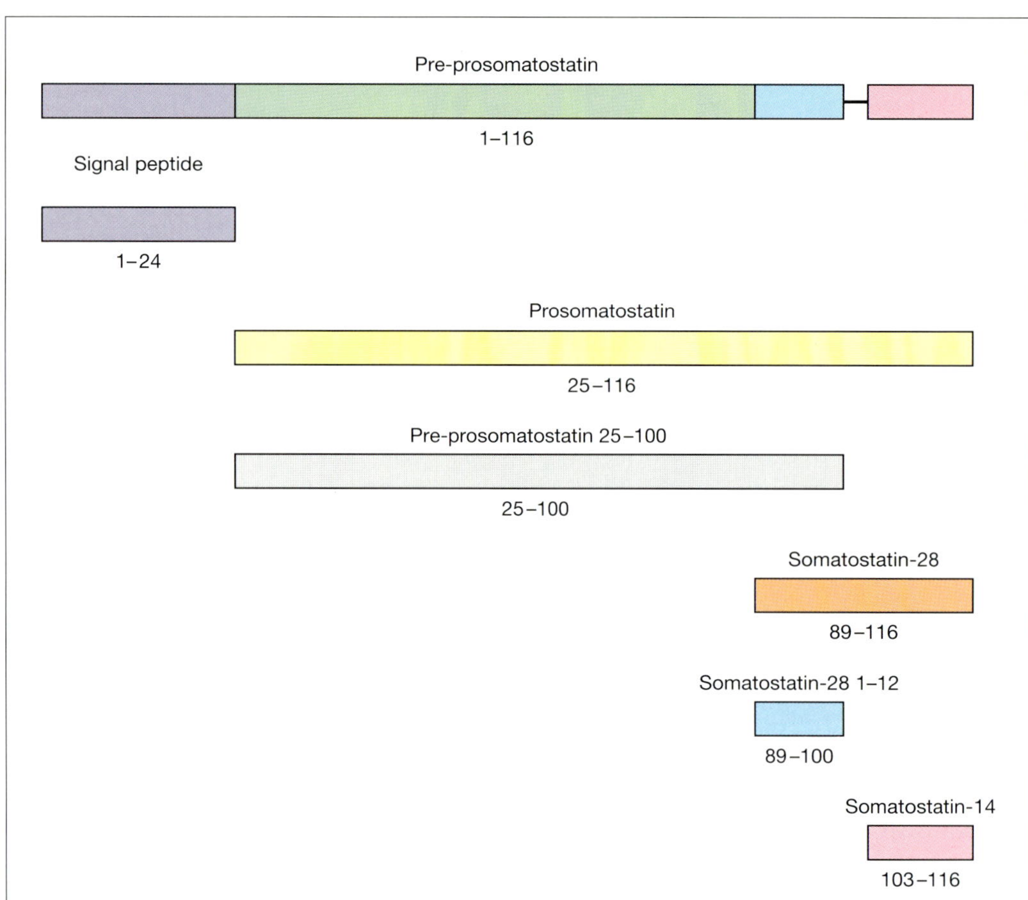

Fig. 4. *SS-14 and related cleavage products of prepro-SS*

coeruleus, and septal nuclei [36]. Postmortem studies in humans have revealed a distribution of SS neurons and fibers similar (although not identical) to that seen in rodents [8]. Several CNS SS pathways have been identified. The periventricular nucleus is the primary source of somatostatinergic input to the median eminence [74], while the SS-containing nerve terminals of the anterior hypothalamic nucleus and lateral and ventromedial hypothalamus originate in the amygdala [3]. Periventricular hypothalamic somatostatinergic fibers travel in a variety of pathways, including short-distance projections to hypothalamic nuclei (e.g., suprachiasmatic, arcuate, and preoptic nuclei), long-distance rostral and ascending projections to many limbic structures (e.g. stria terminalis, amygdala, and arcuate nucleus), and caudal projections into the brain stem and spinal cord [88].

SS receptors are selectively distributed throughout the brain [118]. Several forms of the SS-14 receptor have been identified in human brain that display different localization and affinity characteristics [91, 129], although Moyse et al. [67] have suggested that other factors (e.g., manganese) mediate differences in affinity of a single SS receptor. Finally, the SS tetradecapeptide SS-14 is a C-terminal cleavage product of a larger prohormone and is, therefore, only one of a family of peptides that are identified as "SS-like immunoreactivity" in conventional radioimmunoassays (the antisera of which are frequently directed toward the midportion of the tetradecapeptide). A number of N-terminal extensions of SS-14 have been identified including SS-28, pro-SS (92 amino acids), and prepro-SS (116 amino acids) [9] (Fig. 4). These SS-related peptides appear in the CSF [86] and possess biological activities and potencies as well as patterns of distribution in the CNS that differ from those of SS-14 [118]. The fragments that remain after SS-14 is cleaved from the prohormone (8-kDa fragment) or from SS-28 by convertase [41] ($SS\text{-}28_{1-12}$) have also been identified, and it appears that SS-28, $SS\text{-}28_{1-12}$, and SS-14 are the three most abundant forms of SS in the brain.

While distinct roles for the SS-related peptides in neuropsychiatric disorders cannot be assigned at present, chromatographic studies suggest differential alterations in SS fragments in at least one disorder. Thus, Gomez et al. [42] observed in patients with Alzheimer's disease a decrease in CSF SS-14 and an increase in the concentration of the 15-kDa precursor, thus suggesting the contribution of a post-translational dysfunction to the decreased CSF SS seen in this disorder. Rissler et al. [93], who observed considerable heterogeneity of SS-related peptides in demented patients and controls, also suggested that abnormal processing of SS precursors may be of syndromic relevance. Further study of SS-related peptide profiles in neuropsychiatric disorders is clearly indicated.

Neurophysiological Actions

Both inhibitory and excitatory effects of SS on neuronal activity have been described in electrophysiological studies. These ostensibly contradictory findings may reflect regional specificities as well as dose-response and time-course characteristics. Thus, intracellular recording studies have revealed a biphasic dose-related neuronal response to SS, with low doses increasing and high doses inhibiting, or not affecting, action potential generation [28]. More recent studies have consistently demonstrated the ability of SS to hyperpolarize CA1 pyramidal and solitary tract neurons, with both pharmacological and voltage-clamp evidence for SS-induced m-current (a voltage-dependent outward potassium conductance) activation as the mechanism for this hyperpolarization [49]. Further, this SS-activated potassium conductance appears to be sensitive to pertussis toxin and thus mediated by G protein [90].

Interaction with Neuroregulators

SS secretion is intimately related to the regulation of classical neurotransmitters and neuropeptides. It increases the turnover of cholinergic and monoaminergic neurotransmitters and additionally modulates their release [88, 101]. In turn, SS release is stimulated by dopamine, probably via the D_2 receptor [59], and appears to be stimulated by (muscarinic) acetylcholine [96] and glutamate – via the *N*-methyl-D-aspartate (NMDA) receptor [125] – and inhibited by γ-aminobutyric acid (GABA) [96]. As summarized by Reichlin [88], discrepancies in the reported effects of neurotransmitters on SS may in part reflect differences in the brain regions studied and the methods employed.

The reported effects of hormones on SS content and release are, again, region and method dependent. The hypothalamic content and release of SS are increased by GH [10] and decreased by ACTH [95]. Neurotensin has been reported to stimulate SS secretion from both the hypothalamus and the neocortex, while vasoactive intestinal peptide (VIP) has been observed to inhibit SS secretion from the hypothalamus but to stimulate its release from the neocortex [95].

The effects of SS on the secretion of an array of hormones is so uniformly inhibitory as to have resulted in the suggestion by McCann [63] that SS be called "panhibin." SS appears most clearly involved in the physiological regulation of basal and stimulated GH and thyroid-stimulating hormone (TSH). Tannenbaum and Ling [124] have suggested that the ultradian GH rhythm reflects the tonic secretion of SS and growth-hormone-releasing hormone (GHRH) as well as superimposed rhythmic surges of each at 3- to 4-h intervals, 180° out of phase. The effect of SS on another anterior pituitary hormone, ACTH, is less clear. While Heisler et al. [45] described the ability of SS to inhibit stimulated ACTH secretion following administration of several secretagogues in mouse pituitary tumor (AtT-20) cells, Brown et al. [18] did not observe inhibition by SS of ACTH release stimulated by corticotropin-releasing hormone (CRH) in pituitary cell culture. In this report, however, it was noted that intracerebroventricularly (ICV) administered SS-28 inhibited stress-induced CRF secretion. Additionally, SS infusion in humans was observed by Petraglia et al. [80] to blunt insulin-induced hypoglycemia-stimulated elevations of β-endorphin, β-lipotropin and cortisol. SS appears capable of inhibiting pituitary hormone secretion by at least two mechanisms: (1) inhibition of adenylate cyclase (through G_i protein activation) with consequent reduction in cyclic adenosine monophosphate (cAMP) and (2) activation of a voltage-dependent potassium channel resulting in membrane hyperpolarization, decreased calcium influx, and decreased intracellular free calcium. Reisine et al. [90] have suggested that in AtT-20 cells both cAMP-dependent and cAMP-independent calcium influxes are blocked by pertussis toxin and may, therefore, be mediated by the same SS receptor. However, while SS totally inhibits GHRH-stimulated calcium efflux [60] and GH secretion, it decreases GHRH-stimulated cAMP by only 50% [109], thus suggesting the physiological relevance for the action of SS of the distinction between cAMP-dependent and cAMP-independent processes. Finally, SS has also been reported to inhibit protein phosphorylation [29].

In summary, the evidence in support of the role of SS as a neuromodulator, if not a neurotransmitter, includes its selective concentration in brain regions, neurophysiological actions, localization in synaptic granules in neuronal terminal regions, calcium-dependent release from neuron terminals, degradation by enzymes found in brain extracts, and binding to specific brain receptors [87].

Behavioral Actions

The effects of SS on CNS activity are further demonstrated by the alterations observed in several vegetative and related functions (many of which are also disturbed in depression) following central administration. Intracerebral or ICV administration of SS was observed to produce decreased total and slow-wave sleep as well as decreased rapid eye movement (REM) sleep [43, 92]; however, in a more recent study, Danguir [25] reported a significant increase in REM sleep in rats following ICV SS as well as suppression of REM sleep following ICV administration of the SS depletor cysteamine.

Intracerebral or ICV SS has also been observed to produce increased or dose-related biphasic food consumption. Danguir [26] described a significant increase in daily food intake following chronic ICV administration of the SS analogue octreotide, while SS antiserum significantly decreased food intake. SS has also been shown to antagonize CRH-induced anorexia in starved rats [110]. As reviewed by Vecsei and Widerlov [132], SS influences locomotor activity, the nature of the response being dependent upon the SS analogue, the dose, and the behavioral measure employed.

Intrathecal administration of SS produces analgesia in both animals and man [44, 64]. Early reports of agonist or partial agonist effects of SS at the opiate receptor [85] have gained support from observations of high affinity of several SS analogues at the μ opiate receptor [133]. However, the analgesic effects of intrathecal SS in man are not reversed by naloxone [64], a fact suggesting a nonopiate-mediated mechanism. Bowker [16] has reported the colocalization of SS and methionine enkephalin in neurons of the raphe nucleus and nucleus gigantocellularis. Finally, a series of studies by Vecsei and Widerlov [132] and Walsh et al. [134] have demonstrated that SS may enhance learning and reverse induced learning deficits, while SS depletion (with cysteamine) diminishes performance.

Somatostatin in Affective Disorder: Clinical Relevance

Since, as reviewed elsewhere [87], the brain appears to be a major source of SS in the CSF, the decreased CSF SS seen in depression may directly reflect alterations in brain somatostatinergic function. However, given the existence of several neuropsychiatric disorders that are characterized by decreased CSF SS, its seems likely that SS dysregulation reflects or contributes to clinical or symptomatic aspects of these disorders, rather than playing a unique etiological role in depression. Thus, decreases in SS in depression may occur secondary to a primary central neuroregulatory disturbance or may be epiphenomenal to depression-related diurnal abnormalities [135].

We observed no significant diurnal effect on CSF SS in depressed patients or normal volunteers who underwent lumbar punctures both mornings and evenings. However, in contrast to the differences noted between patient and volunteer SS values from the morning samples, highly similar mean SS values were observed in both groups during the evening [102]. Thus, the detection of decreased CSF SS levels in depressed patients relative to normal volunteers appears to be dependent on the time of day at which the samples are obtained and, therefore, may reflect either a disturbed temporal secretion or a decrease in total SS secretion. Of note in this regard is the report by Sorensen et al. [115] of the loss of diurnal variation as part of the general decrease in CSF SS in patients with MS in relapse compared with patients in the stable phase or controls.

Finally, it is possible that decreased CSF SS, in the disorders characterized by this abnormality, reflects a common central neuroregulatory disturbance or mediates a common manifestation of these disorders (e.g., cognitive impairment or cortisol hypersecretion). For example, as noted above, central SS administration facilitates learning and memory (delays extinction of a learned avoidance response and reverses electric-shock-induced amnesia), and depletion of SS results in impaired performance in the same learning paradigms [132, 134]. These findings are consistent with the observation that cognitive impairment characterizes the neuropsychiatric disorders associated with decreased SS, i.e., Alzheimer's disease, depression, MS, Parkinson's disease, and Cushing's disease. Further, the extent of the decrease in CSF (or brain) SS has been reported by some (but not all) [2, 4, 27, 122] to be significantly correlated with the degree of cognitive impairment in patients with Alzheimer's disease [23, 37, 89, 113, 114] and Parkinson's disease [33].

Endocrine Correlates

In addition to the potential role for SS in the symptoms of depression, alterations in SS function may mediate some of the physiological disturbances described in depression. For example, dysregulation of the hypothalamic-pituitary-adrenal (HPA) axis is the best described biological abnormality in depression. A variety of in vivo and in vitro data suggest the reciprocal regulatory effects of the somatostatinergic and HPA systems. Thus, the reported ability of SS to inhibit stimulated ACTH secretion [45] and/or stress-induced CRH secretion [18] suggests the possibility that decreased CSF SS might contribute to the disinhibition of the HPA axis seen in depression. Consistent with this possibility was the observation by Doran et al. [30] that depressed and schizophrenic patients who escaped from dexamethasone suppression displayed significantly lower CSF SS levels. Further, a significant inverse relationship was observed in depressed patients between the maximum post-dexamethasone plasma cortisol and CSF SS levels (r=–0.71). While this relationship was not observed by Agren and Lundqvist [1] or Bissette et al. [13], a very similar negative correlation (r=–0.66) between CSF SS and response to dexamethasone was reported by Serby et al. [107] in patients with Alzheimer's disease as well as by Kaye et al. [53] (r=–0.76) in underweight anorectics.

Several recent studies suggest that this negative relationship between SS and HPA activity may reflect hypercortisolemia-induced decreases in SS. Wolkowitz et al. [139] reported significant reductions in CSF SS levels in normal volunteers following 5 days of prednisone (80 mg/day) administration. Decreased CSF SS has also been observed in patients with ACTH-dependent Cushing's syndrome [54]. In cell culture, Cote et al. [21] described dramatic reductions in SS messenger ribonucleic acid (mRNA) and SS secretion following the application of dexamethasone.

Even if it should subsequently be demonstrated that depression-related decreases in CSF SS reflect rather than cause HPA axis disinhibition, evidence to date is sufficient to suggest that the relationship between the somatostatinergic and HPA systems is physiologically meaningful and may further our understanding of those disorders characterized by their dysregulation. In addition to the reports described above, this evidence includes the following:

1. Demonstration of CRH-stimulated SS secretion from cultured brain cells [79]
2. Identification of CRH-stimulated SS secretion as the mechanism for stress- and CRH-induced GH suppression (seen in rodents, but not in man) [94]

3. Demonstration of significant reductions in SS receptors in the hippocampus, striatum, and hypothalamus following adrenalectomy, with restoration accompanying replacement with dexamethasone [99]
4. Observation of increased levels of plasma SS following CRH infusion in humans [58]
5. Observation of a significant positive correlation between CSF SS and CRH in a large group of psychiatric patients (n=181) and in controls (n=57) (Rubinow et al., unpublished data)

The further observation that the correlation between CSF SS and CRH existed in controls and in some patient groups (acute and abstinent bulimic [53], euthymic bipolar [12], chronic fatigue syndrome [Demitrack et al., unpublished data], and Cushing's disease patients [54]) but not others (anorexia nervosa [53], schizophrenia) suggests that the absence of a relationship between CSF SS and CRH in certain psychiatric disorders may provide evidence of a dysregulated system.

Role in Seizure Disorders and Effects of Psychopharmacological Agents

While the precise role of SS in depression and other neuropsychiatric disorders is unknown, the potential relevance of alterations in SS secretion to CNS activity and seizure development is increasingly apparent. Supporting evidence includes the following:

1. Long-term (but not permanent) increases in SS in selective brain regions (amygdala and sensorimotor, piriform, and entorhinal cortex; striatum) following kindling (an animal model of epilepsy in which daily subictal stimuli result in progressively more intense brain activity culminating in generalized seizures) or prekindling (rats injected with pentylenetetrazol, PTZ, but not fulfilling the criteria of kindling) [52, 81]
2. Reversible decreases in SS during acute seizures stimulated by PTZ or the neurotoxin kainic acid (and presumed to be related to seizure-related increased release) [81, 117]
3. A 200% increase in SS precursor and 60% and 80% increases in SS-14 and SS-28, respectively, in the frontal cortex after both PTZ-kindled and kainic-acid-stimulated seizures [62]
4. Long-lasting increases (30 days) in prepro-SS mRNA in the frontal cortex and hippocampus (but not striatum or substantia nigra) following kainic-acid-induced seizures [72]
5. Kindling-related downregulation of SS receptors in the hippocampus [46]
6. Periventricular SS release following electrical stimulation of several limbic sites [52]
7. Precipitation of seizures following ICV or intracerebral injection of SS [51]
8. Inhibition of kindled seizures following administration of cysteamine or SS antiserum [47]
9. Reversal of the inhibitory effects of cysteamine by ICV infusion of SS [77]
10. Selective increases in SS and prepro-SS mRNA (but not glutamic acid decarboxylase, GAD, or GAD mRNA) in the striatum and neocortex in stage 5 kindled rats [111]
11. Increases in SS-14 in focal (epileptic) compared with nonfocal (nonepileptic) regions of the temporal cortex in 33 of 35 patients with intractable seizures [69]
12. Increases of SS in stage 3 kindled rats (forelimb clonus) and in stage 5 kindled rats (generalized seizures, rearing and falling), with the elevation in the latter observed 1 h after the seizure and persisting up to 2 weeks following the last seizure [68]

Despite this impressive array of evidence, the findings of several studies have been interpreted as being incompatible with SS having a role in seizure susceptibility or activity. Pitkanen et al. [83] observed no change in CSF SS in humans following a generalized seizure, and Cottrell and Robertson [22] suggested that cysteamine inhibits kindled seizures by inducing myoclonic seizures and not as a consequence of SS depletion (a finding that would appear to be inconsistent with the findings of Perlin et al. [77]).

Nonetheless, additional support for a potential role of SS in seizure disorders is found in reports of the effects on SS of several anticonvulsants. We observed significant reductions in CSF SS levels in affectively ill patients during treatment with the tricyclic anticonvulsant carbamazepine compared with baseline, medication-free values [105]. Patients were selected for lack of major mood state changes during treatment; further, any improvement in depression observed with carbamazepine would have tended to increase rather than decrease SS. Steardo et al. [119] also described a significant decrease in CSF SS in patients with temporal lobe epilepsy during treatment with carbamazepine. In rats, Pitkanen et al. [82] observed lower CSF SS 2 h after intraperitoneal administration of carbamazepine. This effect of carbamazepine on CSF SS seems most likely due to carbamazepine-induced changes in central neuroregulators. While reports suggest the ability of carbamazepine to influence the secretion or action of many of the neurotransmitters involved in SS regulation (dopamine, GABA, acetylcholine, norepinephrine) [84], the effects of carbamazepine on GABA activity appear relevant. As mentioned above, GABA-mimetic agents appear to decrease SS secretion. Carbamazepine decreases GABA turnover, perhaps by a GABA agonist mechanism [11]; additionally, it has been suggested that the action of carbamazepine at the trigeminal nucleus is mediated via baclofen-like or GABA-$_B$ receptor mechanisms [126]. Further, Stryker et al. [120] have shown that SS secretion from diencephalic and cerebral cultures is inhibited by micromolar concentrations of a benzodiazepine, midazolam, an effect that they postulate occurs by stimulation of the GABA-$_B$ receptor.

Finally, Scharfman and Schwartzkroin [106] have recently described the ability of SS to act presynaptically in the CA1 region of the hippocampus and to dramatically depress

GABA-mediated synaptic potentials. As the depression of inhibitory postsynaptic potential (IPSPs) by SS was prolonged, the authors speculated that SS-related inhibition of GABA-induced hyperpolarization might render the hippocampus hyperexcitable and facilitate the induction of long-term potentiation (LTP). Thus, GABA and SS may have reciprocal, inhibitory regulatory effects that, in combination, determine the excitability of brain regions such as the hippocampus. Carbamazepine, then, may act as an anticonvulsant in part by decreasing somatostatinergic activity, a hypothesis that derives support from the report by Higuchi et al. [48] of the blunting by carbamazepine of kindling-induced increases of SS in the frontal, temporal, and occipital cortex and amygdala (but not hippocampus). Additionally, Nagaki et al. [70] recently observed decreased SS and increased GABA levels in several (primarily limbic) brain regions following acute and chronic high-dose carbamazepine administration. It should, nonetheless, be noted that in two other studies the reported effects of carbamazepine on basal SS concentration in the brains of nonkindled rats have been only slight [82] or absent [136]. A direct relationship of decreases in SS to the anticonvulsant effects of carbamazepine remains to be determined.

The effects on SS of several other psychopharmacological agents have been described. Doran et al. [31] described a significant decrease in CSF SS in schizophrenic patients treated with the neuroleptic fluphenazine. Studies with the neuroleptic haloperidol also show an effect on SS, i.e., decreased SS in the striatum and nucleus accumbens (but not the frontoparietal cortex and hippocampus) and decreased SS receptors in the rat cerebral cortex and hippocampus [76]. However, Gattaz et al. [39] reported an increase in CSF SS in a small group of patients during treatment with haloperidol. These studies and others [127] suggest that some of the effects of dopamine blockers may be mediated by alterations in SS. Finally, no significant effects on CSF SS were noted during treatment with several other psychopharmacological agents including desmethylimipramine, piribedil, and lithium carbonate [12, 104]. Kakigi et al. [50] similarly reported no effects of imipramine, maprotiline, or mianserin on SS concentration in rat brain; they did, however, observe significant decreases in SS concentrations in a variety of brain regions following administration of serotonergic agents (clomipramine, zimelidine, 5-hydroxytryptophan).

Conclusions

Decreases in brain and/or CSF SS have been identified and replicated in many neuropsychiatric disorders, including Alzheimer's disease, depression, MS in relapse, Parkinson's disease, and Cushing's disease. While the decreases in brain and CSF SS in Alzheimer's disease appear to be the product of a degenerative process, the SS dysregulation seen in depression, like that in MS, is transient and reversible and appears to reflect functional alterations in SS secretion and/or metabolism. Changes in SS appear intimately linked to the cognitive decrements in Alzheimer's disease, raising the possibility that cognitive impairments in the other neuropsychiatric syndromes characterized by low SS could be similarly linked. Promising areas for further investigation also include the relationship of low SS to mood, motor, convulsive, and endocrine disturbances of these major neuropsychiatric disorders. Clarification of the precise mechanisms involved in the SS alterations and their consequences should provide valuable insights into normal and pathological CNS function.

References

1. Agren H, Lundqvist G (1984) Low levels of somatostatin in human CSF mark depressive episodes. Psychoneuroendocrinology 9: 233–248
2. Atack JR, Beal MF, May C, Kaye JA, Mazurek MF, Kay AD, Rapoport SI (1988) Cerebrospinal fluid somatostatin and neuropeptide Y: concentrations in aging and in dementia of the Alzheimer type without extrapyramidal signs. Arch Neurol 45: 269–274
3. Beal MF, Domesick VB, Martin JB (1985) Effects of lesions in the amygdala and periventricular hypothalamus on striatal somatostatin-like immunoreactivity. Brain Res 330: 309–316
4. Beal MF, Growdon JH, Mazurek MF, Martin JB (1986) CSF somatostatin-like immunoreactivity in dementia. Neurology 36: 294–297
5. Beal MF, Mazurek MF, Black PMcL, Martin JB (1985) Human cerebrospinal fluid somatostatin in neurologic disease. J Neurol Sci 71: 91–104
6. Beal MF, Mazurek MF, Ellison DW, Swartz KJ, McGarvey U, Bird ED, Martin JB (1988) Somatostatin and neuropeptide Y concentrations in pathologically graded cases of Huntington's disease. Ann Neurol 23: 562–569
7. Beal MF, Mazurek MF, Svendsen CN, Bird ED, Martin JB (1986) Widespread reduction of somatostatin-like immunoreactivity in the cerebral cortex in Alzheimer's disease. Ann Neurol 20: 489–495
8. Bennett-Clarke CA, Joseph SA (1986) Immunocytochemical localization of somatostatin in human brain. Peptides 7: 877–884

9. Benoit R, Bohlen P, Esch F, Ling N (1984) Neuropeptides derived from prosomatostatin that do not contain the somatostatin-14 sequence. Brain Res 311: 23–29
10. Berelowitz M, Firestone SL, Frohman LA (1981) Effects of growth hormone excess and deficiency on hypothalamic somatostatin content and release and on tissue somatostatin distribution. Endocrinology 109: 714
11. Bernasconi R, Martin P (1979) Effects of antileptic drugs on the GABA turnover rate. Naunyn Schmiedebergs Arch Pharmacol 307: R63
12. Berrettini WH, Nurnberger JI Jr, Zerbe RL, Gold PW, Chrousos GP, Tomai T (1987) CSF neuropeptides in euthymic bipolar patients and controls. Br J Psychiatry 150: 208–212
13. Bissette G, Widerlov E, Walleus H, Karlsson I, Eklund K, Forsman A, Nemeroff CB (1986) Alterations in cerebrospinal fluid concentrations of somatostatin-like immunoreactivity in neuropsychiatric disorders. Arch Gen Psychiatry 43: 1148–1151
14. Black PM, Ballantine HT, Carr DB, Beal MF, Martin JB (1986) Beta-endorphin and somatostatin concentrations in the ventricular cerebrospinal fluid of patients with affective disorder. Biol Psychiatry 21: 1077–1081
15. Bowen DM, Najlerahim A, Procter AW, Francis PT, Murphy E (1989) Circumscribed changes of the cerebral cortex in neuropsychiatric disorders of later life. Proc Natl Acad Sci USA 86: 9504–9508
16. Bowker RM (1987) Evidence for the co-localization of somatostatin- and methionine-enkephalin-like immunoreactivity in raphe and gigantocellularis nuclei. Neurosci Lett 81: 75–81
17. Brazeau P, Vale W, Burgus R, Ling N (1973) Hypothalamic polypeptide that inhibits the secretion of immunoreactive pituitary growth hormone. Science 197: 77–79
18. Brown MR, Rivier C, Vale W (1984) Central nervous system regulation of adrenocorticotropin secretion: role of somatostatin: Endocrinology 114: 1546–1549
19. Candy JM, Gascoigne AD, Biggins A, Smith AI, Perry RH, Perry EK, McDermott JR, Edwardson JA (1985) Somatostatin immunoreactivity in cortical and some subcortical regions in Alzheimer's disease. J Neurol Sci 71: 315–323
20. Charlton BG, Leake A, Wright C, Fairbairn AF, McKeith IG, Candy JM, Ferrier IN (1988) Somatostatin content and receptors in the cerebral cortex of depressed and control subjects. J Neurol Neurosurg Psychiatry 51: 719–721
21. Cote GJ, Palmer WN, Leonhart K, Leong SS, Gagel RF (1986) The regulation of somatostatin production in human medullary thyroid carcinoma cells by dexamethasone. J Biol Chem 261: 12930–12935
22. Cottrell GA, Robertson HA (1987) Prevention of cysteamine-induced myoclonus blocks the long-term inhibition of kindled seizures. Brain Res 412: 161–164
23. Cramer H, Schaudt D, Rissler K, Strubel D, Warter JM, Kuntzmann F (1985) Somatostatin-like immunoreactivity and substance-P-like immunoreactivity in the CSF of patients with senile dementia of Alzheimer type, multi-infarct syndrome and communicating hydrocephalus. J Neurol 232: 346–351
24. Cutler NR, Haxby JV, Narang PR (1985) Evaluation of an analogue of somatostatin (L363; 586) in Alzheimer's disease. N Engl J Med 312: 725
25. Danguir (1986) Intracerebroventricular infusion of somatostatin selectively increases paradoxical sleep in rats. Brain Res 367: 26–30
26. Danguir J (1988) Food intake in rats is increased by intracerebroventricular infusion of the somatostatin analogue SMS 201–995 and is decreased by somatostatin antiserum. Peptides 9: 211–213
27. Davis KL, Davidson M, Yang RK, Davis BM, Siever LJ, Mohs RC, Ryan T, Coccaro E, Bierer L, Targum SD (1988) CSF somatostatin in Alzheimer's disease, depressed patients, and control subjects. Biol Psychiatry 24: 710–712
28. Delfs JR, Dichter MA (1983) Effects of somatostatin on mammalian cortical neurons in culture: physiological actions and unusual dose response characteristics. J Neurosci 3: 1176–1188
29. Dokas LA, Kliss M, Liauv A, Coy DA (1983) Evidence that (D-Trp^8) somatostatin inhibits synaptic plasma membrane protein phosphorylation by interaction with a specific membrane binding site. Abstr Soc Neurosci 9: 589
30. Doran A, Rubinow DR, Roy A, Pickar D (1986) CSF somatostatin and abnormal response to dexamethasone administration in schizophrenic and depressed patients. Arch Gen Psychiatry 43: 365–369
31. Doran AR, Rubinow DR, Wolkowitz OM, Roy A, Breier A, Pickar D (1989) Fluphenazine treatment reduces CSF somatostatin in patients with schizophrenia: correlations with CSF HVA. Biol Psychiatry 25: 431–439
32. Dupont E, Christensen SE, Hansen AP, Olivarius BD, Orskov H (1982) Low cerebrospinal fluid somatostatin in Parkinson's disease: an irreversible abnormality. Neurology (NY) 32: 312–314
33. Epelbaum J, Ruberg M, Moyse E, Javoy-Agid F, Dubois B, Agid Y (1983) Somatostatin and dementia in Parkinson's disease. Brain Res 278: 376–379
34. Ferrier IN, Croxx AJ, Johnson JA, Roberts GW, Crow TJ, Corsellis JA, Lee YC, O'Shaughnessy D, Adrian TE, Mc Gregor GP, Baracese-Hamilton AJ, Bloom SR (1983) Neuropeptides in Alzheimer type dementia. J Neurol Sci 62: 159–170
35. Ferrier IN, McKeith IG, Charlton BG, Whitford CA, Marshall E, Perry RH, Eccleston D, Edwardson JA (1989) Postmortem investigations of serotonergic and peptidergic hypotheses of affective illness. In: Lerer B, Gershon S (eds) New directions in affective disorders. Springer, Berlin Heidelberg New York, pp 60–63
36. Finley JCW, Maderdrut JL, Roger LJ, Petrusz P (1981) The immunocytochemical localization of somatostatin-containing neurons in the rat central nervous system. Neuroscience 6: 2173–2192
37. Francis PT, Bowen DM, Lowe SL, Neary D, Mann DMA, Snowden JS (1987) Somatostatin content and release measured in cerebral biopsies from demented patients. J Neurol Sci 78: 1–16
38. Francis PT, Bowen DM, Neary D, Palo J, Wikstrom J, Olney N (1984) Somatostatin-like immunoreactivity in lumbar cerebrospinal fluid from neurohistologically examined demented patients. Neurobiol Aging 5: 183–186
39. Gattaz WF, Rissler K, Gattaz D, Cramer H (1986) Effects of haloperidol on somatostatin-like immunoreactivity in the CSF of schizophrenic patients. Psychiatry Res 17: 1–6
40. Gerner RH, Yamada T (1982) Altered neuropeptide concentrations in cerebrospinal fluid of psychiatric patients. Brain Res 238: 298–302
41. Glushankof P, Morel A, Gomez S, Nicolas P, Fahy C, Cohen P (1984) Enzyme processing somatostatin precursors: an Arg-Lys esteropeptidase from the rat brain cortex converting somatostatin-28 into somatostatin-14. Proc Natl Acad Sci USA 81: 6662–6666
42. Gomez S, Davous P, Rondot P, Faivre-Bauman A, Valade D, Puymirat J (1986) Somatostatin-like immunoreactivity and acetylcholinesterase activities in cerebrospinal fluid of patients with Alzheimer disease and senile dementia of the Alzheimer type. Psychoneuroendocrinology 11: 69–73
43. Havlicek V, Rezek M, Friesen H (1976) Somatostatin and thyrotropin releasing hormone: central effect of sleep and motor system. Pharmacol Biochem Behav 4: 455–459
44. Havlicek V, Rezek M, Leybin L, Friesen H (1977) Analgesic effect of cerebral ventricular administration of somatostatin. Fed Proc 36: 363
45. Heisler S, Reisine T, Hook V, Axelrod J (1982) Somatostatin inhibits multireceptor stimulation of cyclic AMP formation and corticotrophin secretion in mouse pituitary tumor cells. Proc Natl Acad Sci USA 79: 6502–6507
46. Higuchi T, Kokubu T, Sikand GS, Wada JA, Friesen HG (1984) A study of somatostatin receptors in amygdaloid-kindled rat brain. J Neurochem 43: 1271–1276
47. Higuchi T, Sikand GS, Kato N, Wada JA, Friesen HG (1983) Profound suppression of kindled seizures by cysteamine: possible role of somatostatin to kindled seizures. Brain Res 288: 359–362
48. Higuchi T, Yamazaki O, Takazawa A, Kato N, Watanabe N, Minatogawa Y, Yamazaki J, Ohshima H, Nagaki S, Igarashi Y, Noguchi T (1986) Effects of carbamazepine and valproic acid on brain immunoreactive somatostatin and gamma-aminobutyric acid in amygdala-kindled rats. Eur J Pharmacol 125: 169–175
49. Jacquin T, Champagnat J, Madamba S, Denavit-Saubie M, Siggins

GR (1988) Somatostatin depresses excitability in neurons of the solitary tract complex through hyperpolarization and augmentation of I_m, a non-inactivating voltage-dependent outward current blocked by muscarinic agonists. Proc Natl Acad Sci USA 85: 948–952
50. Kakigi T, Kaneda H, Kawata E, Komurasaki Y, Tanimoto K, Maeda K, Chihara K (1990) The effects of antidepressants and serotonergic agents on the concentration of somatostatin in rat brain (Abstr.) 17th Congress of Collegium Internationale Neuro-Psychopharmacologicum, Kyoto
51. Katakami H, Arimura A, Frohman LA (1985) Involvement of hypothalamic somatostatin in the suppression of growth hormone secretion by central corticotropin-releasing factor in conscious male rats. Neuroendocrinology 41: 390–393
52. Kato M, Kakegawa T, Suzuki M (1986) Electrical stimulation of several limbic areas suppressed the growth hormone (GH) release induced by human pancreatic GH-releasing factor in pentobarbital-anesthetized rats. Endocrinology 118: 955–960
53. Kaye W, Rubinow DR, Gwirtsman H, Tomai T, Chrousos G, Gold PW (1988) CSF somatostatin in anorexia nervosa and bulimia: relationship to the hypothalamic-pituitary-adrenal cortical axis. Psychoneuroendocrinology 13: 265–272
54. Kling MA, Rubinow DR, Chrousos GP, Doran AR, Gold PW (1986) CSF somatostatin levels are decreased in patients with ACTH-dependent Cushing's syndrome (Abstr. 92) 41st Annual meeting of the Society of Biological Psychiatry
55. Koponen H, Stenback U, Mattila E, Reinikainen K, Soininen H, Riekkinen PJ (1989) Cerebrospinal fluid somatostatin in delirium. Psychol Med 19: 605–609
56. Kruesi MJP, Swedo S, Leonard H, Rubinow DR, Rapoport JL (1990) CSF somatostatin in childhood psychiatric disorders: a preliminary investigation. Psychiatry Res 33: 277–284
57. Krulich L, Dhariwal APS, McCann SM (1968) Stimulatory and inhibitory effects of purified hypothalamic extracts on growth hormone release from rat pituitary in vitro. Endocrinology 83: 783–790
58. Lesch KP, Widerlov E, Ekman R, Laux G, Rupprecht R, Schulte HM, Beckmann H (1989) The influence of human corticotropin-releasing hormone on somatostatin secretion in depressed patients and controls. J Neural Transm 75: 111–118
59. Lewis BM, Diegues C, Lewis M, Hall R, Scanlon MF (1986) Hypothalamic D2 receptors mediate the preferential release of somatostatin-28 in response to dopaminergic stimulation. Endocrinology 119: 1712–1717
60. Login IW, Judd AM (1986) Trophic effects of somatostatin on calcium flux: dynamic analysis and correlation with pituitary hormone release. Endocrinology 119: 1703–1707
61. Lowe SL, Francis PT, Procter AW, Palmer AM, Davison AN, Bowen DM (1988) Gamma-aminobutyric acid concentration in brain tissue at two stages of Alzheimer's disease. Brain 111: 785–799
62. Marksteiner J, Sperk G (1988) Concomitant increase of somatostatin, neuropeptide Y and glutamate decarboxylase in the frontal cortex of rats with decreased seizure threshold. Neuroscience 26: 379–385
63. McCann SM (1982) Physiology and phamacology of LHRH and somatostatin. Annu Rev Pharmacol Toxicol 22: 491–515
64. Meynadier J, Chrubasik J, Dubar M, Wunsch E (1985) Intrathecal somatostatin in terminally ill patients. A report of two cases. Pain 23: 9–12
65. Molchan SE, Lawlor BA, Hill JL, Martinez RA, Davis CL, Mellow AM, Rubinow DR, Sunderland T (1991) CSF monoamine metabolites and somatostatin in Alzheimer's disease and major depression. Biol Psychiatry 29: 1110–1118
66. Morrison JH, Rogers J, Scherr S, Benoit R, Bloom FE (1985) Somatostatin immunoreactivity in neuritic plaques of Alzheimer's patients. Nature 314: 90–92
67. Moyse E, Slama A, Videau C, DeAngela P, Kordon C, Epelbaum J (1989) Regional distribution of somatostatin receptor affinity states in rat brain: Effects of divalent cations and GTP. Regul Pept 26: 225–234
68. Nadi NS (1988) The time course of chemical changes in the kindled rat brain. Abstr Soc Neurosci 14: 1031
69. Nadi NS, Holmes MD, Wyler AR, Porter RJ (1986) Neuropeptides in focal and nonfocal epileptic tissues from patients with intractable epilepsy. Abstr Soc Neurosci 12: 294
70. Nagaki S, Kato N, Minatogawa Y, Higuchi T (1990) Effects of anticonvulsants and gamma-aminobutric acid (GABA)-mimetic drugs on immunoreactive somatostatin and GABA contents in the rat brain. Life Sci 46: 1587–1595
71. Nemeroff CB, Youngblood WW, Manberg PJ, Prange AJ, Kizer JS (1983) Regional brain concentrations of neuropeptides in Huntington's chorea and schizophrenia. Science 221: 972–975
72. Olenik C, Meyer DK, Marksteiner J, Sperk G (1989) Concentrations of mRNAs encoding for preprosomatostatin and preprocholecystokinin are increased after kainic acid-induced seizures. Synapse 4: 223–228
73. Oram JJ, Edwardson J, Millard PH (1981) Investigation of cerebrospinal fluid neuropeptides in idiopathic senile dementia. Gerontology 27: 216–223
74. Palkovits M, Tapia-Arancibia L, Kordon C, Epelbaum J (1982) Somatostatin connections between the hypothalamus and the limbic system of the rat brain. Brain Res 250: 223–228
75. Patel YC, Rao K, Reichlin S (1977) Somatostatin in human cerebrospinal fluid. N Engl J Med 296: 529–533
76. Perez-Oso E, Lopez-Ruiz MP, Arilla E (1990) Effect of haloperidol withdrawal on somatostatin level and binding in rat brain. Biosci Rep 10: 15–22
77. Perlin JB, Lothman EW, Geary WA (1987) Somatostatin augments the spread of limbic seizures from the hippocampus. Ann Neurol 21: 475–480
78. Perry EK, Blessed G, Tomlinson BE, Perry RH, Crow TJ, Cross AJ, Dockray GJ, Dimaline R, Arregui A (1981) Neurochemical activities in human temporal lobe related to aging and Alzheimer-type changes. Neurobiol Aging 2: 251–256
79. Peterfreund RA, Vale WW (1983) Ovine corticotropin-releasing factor stimulates somatostatin secretion from cultured brain cells. Endocrinology 112: 1275–1278
80. Petraglia F, Facchinetti F, D'Ambrogio G, Volpe A, Genazzani AR (1986) Somatostatin and oxytocin infusion inhibits the rise of plasma B-endorphin, B-lipotrophin and cortisol induced by insulin hypoglycaemia. Clin Endocrinol (Oxf) 24: 609–616
81. Pitkanen A, Beal MF, Riekkinen PJ (1988) Levels of somatostatin and neuropeptide Y in pentylenetetrazol-kindled rat brain. Epilepsia 29: 659
82. Pitkanen A, Hyttinen JM, Sirvio J, Jokkonen J, Sarlund H, Riekkinen P (1989) The levels of somatostatin in the brain and CSF of rat after carbamazepine administration. Pharmacol Toxicol 64: 266–271
83. Pitkanen A, Jolkkonen J, Riekkinen P (1987) B-endorphin, somatostatin, and prolactin levels in cerebrospinal fluid of epileptic patients after generalised convulsion. J Neurol Neurosurg Psychiatry 50: 1294–1297
84. Post RM, Ballenger JC, Uhde TW, Bunney WE (1984) Efficacy of carbamazepine in manic-depressive illness: implications for underlying mechanisms. In: Post RM, Ballenger JC (eds) Neurobiology of mood disorders, Williams and Wilkins, Baltimore, pp 777–816
85. Pugsley TA, Lipmann W (1978) Effect of somatostatin analogues and 17-alphadihydroequilin on rat brain opiate receptors. Res Commun Chem Pathol Pharmacol 21: 153–156
86. Reichlin S (1982) Somatostatin in the nervous system. In: Schmitt FO, Bird SJ, Blood FE (eds) Molecular genetic neuroscience. Raven, New York, pp 359–372
87. Reichlin S (1983) Somatostatin. In: Krieger DT, Brownstein MJ, Martin JB (eds) Brain peptides, Wiley, New York, pp 711–752
88. Reichlin S (1983) Somatostatin. N Engl J Med 309: 1556–1563
89. Reinikainen KJ, Riekkinen PJ, Jolkkonen J, Kosma VM, Soininen H (1987) Decreased somatostatin-like immunoreactivity in cerebral cortex and cerebrospinal fluid in Alzheimer's disease. Brain Res 402: 103–108
90. Reisine T, Wang HL, Guild S (1988) Somatostatin inhibits cAMP-dependent and cAMP -independent calcium influx in the clonal pituitary tumor cell line AtT-20 through the same receptor population. J Pharmacol Exp Ther 245: 225–231
91. Reubi JC, Probst A, Cortes R, Palacios JM (1987) Distinct topo-

graphical localisation of two somatostatin receptor subpopulations in the human cortex. Brain Res 406: 391–396

92. Rezek M, Havlicek V, Hughes KR, Friesen H (1976) Central site of action of somatostatin (SRIF): role of hippocampus. Neuropharmacology 15: 499–504
93. Rissler K, Cramer H, Schaudt D, Strubel D, Gattaz WF (1986) Molecular size distribution of somatostatin-like immunoreactivity in the cerebrospinal fluid of patients with degenerative brain disease. Neurosci Res 3: 213–225
94. Rivier C, Vale W (1985) Involvement of corticotropin-releasing factor and somatostatin in stress-induced inhibition of growth hormone secretion in the rat. Endocrinology 117: 2478–2482
95. Robbins RJ (1987) Regulation of hypothalamic somatostatin secretion. In: Reichlin S (ed) Somatostatin: basic and clinical status. Plenum, New York, pp 149–156
96. Robbins RJ, Sutton RE, Reichlin S (1982) Effects of neurotransmitters and cylic-AMP on somatostatin release from cultured cerebral cortical cells. Brain Res 243: 377–386
97. Roberts GW, Crow TJ, Polak JM (1985) Localization of neuronal tangles in somatostatin neurons in Alzheimer's disease. Nature 314: 92–94
98. Roberts GW, Ferrier IN, Lee Y, Crow TJ, Johnstone EC, Owens DGC, Bacarese-Hamilton AJ, McGregor G, O'Shaughnessey D, Polak JM, Bloom SR (1983) Peptides, the limbic lobe and schizophrenia. Brain Res 288: 199–211
99. Rodriguez MN, Gomez-Pan A, Arilla E (1988) Decrease in number of somatostatin receptors in rat brain after adrenalectomy: normalization after glucocorticoid replacement. Endocrinology 123: 1147–1152
100. Rosler N, Reuner C, Geiger J, Rissler K, Cramer H (1990) Cerebrospinal fluid levels of immunoreactive substance P and somatostatin in patients with multiple sclerosis and inflammatory CNS disease. Peptides 11: 181–183
101. Rossor MN, Emson PC, Mountjoy CQ, Roth M, Iversen LL (1980) Reduced amounts of immunoreactive somatostatin in the temporal cortex in senile dementia of Alzheimer type. Neurosci Lett 20: 373–377
102. Rubinow DR (1986) Cerebrospinal fluid somatostatin and psychiatric illness. Biol Psychiatry 21: 341–365
103. Rubinow DR, Gold PW, Post RM, Ballenger JC, Cowdry R, Bollinger J, Reichlin S (1983) CSF somatostatin in affective illness. Arch Gen Psychiatry 40: 409–412
104. Rubinow DR, Gold PW, Post RM, Ballenger JC, Reichlin S (1983) Cerebrospinal fluid somatostatin in primary affective disorder. Psychopharmacol Bull 19: 422–425
105. Rubinow DR, Post RM, Gold PW, Ballenger JC, Reichlin S (1985) Effects of carbamazepine on cerebrospinal fluid somatostatin. Psychopharmacology (Berlin) 85: 210–213
106. Scharfman HE, Schwartzkroin PA (1989) Selective depression of GABA-mediated IPSPs by somatostatin in area CA1 of rabbit hippocampal slices. Brain Res 493: 205–211
107. Serby M, Richardson SB, Rypma B, Twente S, Rotrosen JP 1986) Somatostatin regulation of the CRF-ACTH-cortisol axis. Biol Psychiatry 21: 971–974
108. Serby M, Richardson SB, Twente S, Siekierski J, Corwin J, Rotrosen J (1984) CSF somatostatin in Alzheimer's disease. Neurobiol Aging 5: 187–189
109. Sheppard MS, Moor B, Kraicer J (1985) Release of growth hormone (GH) from purified somatotrophs: interaction of GH-releasing factor and somatostatin and role of adenosine 3′, 5′-monophosphate. Endocrinology 117: 2364–2374
110. Shibasaki T, Kim YS, Yamauchi N, Masuda A, Imaki T, Hotta M, Demura H, Wakabayashi I, Ling N, Shizume S (1988) Antagonistic effect of somatostatin on corticotropin-releasing factor-induced anorexia in the rat. Life Sci 42: 329–334
111. Shinoda H, Schwartz JP, Nadi NS (1989) Amygdaloid kindling of rats increases preprosomatostatin mRNA and somatostatin without affecting glutamic acid decarboxylase (GAD) mRNA or GAD. Mol Brain Res 5: 243–246
112. Shults C, Steardo L, Barone P (1986) Huntington's disease: effect of cysteamine, a somatostatin-depleting agent. Neurology 36: 1099–1102
113. Soininen HS, Jolkkonen JT, Reinikainen JK, Halonen TO, Riekkinen PJ (1984) Reduced cholinesterase activity and somatostatin-like immunoreactivity in the cerebrospinal fluid of patients with dementia of the Alzheimer type. J Neurol Sci 63: 167–172
114. Soininen HS, Riekkinen PJ, Partanen J, Helkala EL, Laulumaa V, Jolkkonen J, Reinikainen K (1988) Cerebrospinal fluid somatostatin correlates with spectral EEG variables and with parietotemporal cognitive dysfunction in Alzheimer patients. Neurosci Lett 85: 131–136
115. Sorensen KV, Alslev T, Christensen SE (1987) CSF somatostatin in multiple sclerosis: reversible loss of diurnal oscillation in relapses. Neurology 37: 1050–1053
116. Sorensen KV, Christensen SE, Dupont E, Hansen AP, Pedersen E, Orskov H (1980) Low somatostatin content in cerebrospinal fluid in multiple sclerosis. Acta Neurol Scand 61: 186–191
117. Sperk G, Reynolds GP, Riederer P (1987) HPLC analysis of somatostatin related peptides in putamen of Huntington's disease patients. J Neural Transm 69: 153–160
118. Srikant CB, Patel YC (1985) Somatostatin receptors. Adv Exp Med Biol 188: 291–304
119. Steardo L, Barone P, Hunnicutt E (1986) Carbamazepine lowering effect on CSF somatostatin-like immunoreactivity in temporal lobe epileptics. Acta Neurol Scand 74: 140–144
120. Stryker T, Conlin T, Reichlin S (1986) Influence of a bendzodiazepine, midazolam, and gamma-aminobutyric acid (GABA) on basal somatostatin secretion from cerebral and diencephalic neurons in dispersed cell culture. Brain Res 362: 339–343
121. Su T, Hauser P, Rubinow DR, Elpern S, Maloni HW, Krebs HM, Devinsky O, Post RM, McFarland HF (1990) CSF somatostatin (SRIF), mood, and cognition in multiple sclerosis (Abstr). 21st Congress International Society of Psychoneuroendocrinology, Buffalo, NY
122. Sunderland T, Rubinow DR, Tariot P, Cohen RM, Newhouse PA, Mellow AM, Mueller EA, Murphy DL (1987) CSF somatostatin in dementia, elderly depression and age-matched controls. Am J Psychiatry 144: 1313–1316
123. Tamminga CA, Foster NL, Fedio P, Bird ED, Chase TN (1987) Alzheimer's disease: low cerebral somatostatin levels correlated with impaired cognitive function and cortical metabolism. Neurology 37: 161–165
124. Tannenbaum G, Ling N (1984) The interrelationships of growth hormone (GH)-releasing factor and somatostatin in generation of the ultradian rhythm of GH secretion. Endocrinology 115: 1952–1957
125. Tapia-Arancibia L, Astier H (1989) Actions of excitatory amino acids on somatostatin release from cortical neurons in primary cultures. J Neurochem 53: 1134–1141
126. Terrence CF, Sax M, Fromm GH, Chang CH, Yoo CS (1983) Effect of baclofen enantiomorphs on the spinal trigeminal nucleus and steric similarities of carbamazepine. Pharmacology 27: 85–94
127. Thal LJ, Laing K, Horowitz SG, Makman MH (1986) Dopamine stimulates rat cortical somatostatin release. Brain Res 372: 205–209
128. Thal LJ, Masur DM , Fuld PA, Sharpless NS, Davies P (1983) Memory improvement with oral physostigmine and lecithin in Alzheimer's disease. In: Katzman R (ed) Biological aspects of Alzheimer's disease, vol 15. Cold Spring Harbor Laboratory, New York, pp 461–469
129. Tran VT, Beal MF, Martin JB (1985) Two types of somatostatin receptors differentiated by cyclic somatostatin analogs. Science 288: 492–495
130. Unger J, Weindl A, Ochs G, Struppler A (1988) CSF somatostatin is elevated in patients with postzoster neuralgia. Neurology 38: 1423–1427
131. Vecsei L, Widerlov E (1988) Brain and CSF somatostatin concentrations in patients with psychiatric or neurological illness: an overview. Acta Psychiatr Scand 78: 657
132. Vecsei L, Widerlov E (1990) Effects of somatostatin-28 and some of its fragments and analogs on open-field behavior, barrel rotation, and shuttle box learning in rats. Psychoneuroendocrinology 15: 139–145

133. Walker JM, Bowen WD, Atkins ST, Hemstreet MK, Coy DH (1987) μ-Opiate binding and morphine antagonism by octapeptide analogs of somatostatin. Peptides 8: 869–875
134. Walsh TJ, Emerich DF, Winokur A, Banki C, Bissette G, Nemeroff CB (1985) Intrahippocampal injection of cysteamine depletes somatostatin and produces cognitive impairments in the rat. Neurosci Abstr 11: 621
135. Wehr TA (1984) Biological rhythms and manic-depressive illness. In: Post RM, Ballenger JC (eds) Neurobiology of mood disorders. Williams and Wilkins, Baltimore, pp 190–206
136. Weiss SRB, Nguyen T, Rubinow DR, Helke CJ, Narang PK, Post RM, Jacobowitz DM (1987) Lack of effect of chronic carbamazepine on brain somatostatin in the rat. J Neural Transm 68: 325–333
137. Widerlov E, Bissette G, Nemeroff CB (1988) Monoamine metabolites, corticotropin releasing factor and somatostatin as CSF markers in depressed patients. J Affective Disord 14: 99–107
138. Widerlov E, Brane G, Ekman R, Kihlgren M, Norberg A, Karlsson I (1989) Elevated CSF somatostatin concentrations in demented patients parallel improved psychomotor functions induced by integrity-promoting care. Acta Psychiatr Scand 79: 41–47
139. Wolkowitz OM, Rubinow DR, Breier A, Doran AR, Davis C, Pickar D (1987) Prednisone decreases CSF somatostatin in healthy humans: implications for neuropsychiatric illness. Life Sci 41: 1929–1933
140. Wood PL, Etienne P, Lal S, Gauthier S, Cajal S, Nair NPV (1982) Reduced lumbar CSF somatostatin in Alzheimer's disease. Life Sci 31: 2073–2079

The Physiological Role of Somatostatin in the Regulation of Nutrient Homeostasis

V. Schusdziarra

Department of Internal Medicine II, Technical University of Munich, Klinikum rechts der Isar, W-8000 Munich 80, Federal Republic of Germany

Introduction

The ingestion of food is necessary to cover the basic requirements of the body. The foodstuffs must not only cover the daily energy expenditure but also provide the substrates necessary for the functioning, structural maintenance and growth of the tissues and organs. The gastrointestinal (GI) tract processes the complex foodstuffs in order that nutrients may pass the intestinal wall and be transported in the blood to all parts of the body.

An uncontrolled uptake of nutrients from the intestine into the blood would have harmful consequences on the homeostasis of the *milieu intérieur,* all the more so as the composition of the ingested food may vary considerably from day to day. If the various nutrients were allowed to enter the circulation freely, abnormally high and prolonged concentrations in the blood would arise, and these in turn would result in unphysiological intracellular concentrations of nutrients, the activation of otherwise inactive metabolic pathways and the accumulation of toxic end-products causing severe metabolic disturbances such as gout or the formation of gallbladder or kidney stones. Prolonged hyperglycaemia and hyperlipidaemia have to be considered risk factors for cardiovascular disease. It is therefore necessary that the nutrient flux be regulated in such a manner that the uptake of substrates from the gut into the circulation balances their efflux into the tissues and cells [98]. The achievement of such a balance requires the harmonious interplay of stimulators and inhibitors of digestion.

Among these inhibitors, somatostatin (SS) is of prime importance. It regulates the endocrine and exocrine functions of the stomach, the intestine and the pancreas and the motor functions of the first two organs, partly by local or paracrine mechanisms and partly via the circulation as a true endocrine factor [97, 128]. Accordingly, not only does SS exert its regulatory actions within the organs in which SS-containing D cells are located, but it also ensures a more integrative control over the nutrient flux by acting on remote target organs containing no D cells.

This review will deal with (a) the regulation of SS release during the course of nutrient assimilation and (b) the biological effects of SS on digestive functions that prevent sharp changes from occurring in the blood nutrient levels and in the nutrient flux from the gut to the tissues.

Release of Somatostatin

Basal Somatostatin Release

SS-like immunoreactivity (SLI) has been found in the endocrine-like D cells of the stomach, small intestine and pancreas [4, 21, 36, 56, 69, 76, 78] and in vagal nerve fibres and neurons of the myenteric and submucosal plexuses of the intestinal tract [37, 93, 131]. Within the GI tract, an accumulation of D cells has been demonstrated in the fundic and antral areas of the stomach and in the islets of Langerhans, whereas a more scattered distribution has been shown in the small and large intestine (Fig. 1).

Basal plasma SS levels in the peripheral circulation are the result of SS secretion from the splanchnic organs. In dogs, the highest basal levels of SS have been found in the gastric and pancreatic veins, followed by the portal vein, arteries and the inferior vena cava. Similar results with regard to portoarterial and arteriovenous gradients have been obtained in rats and, to a certain extent, in humans [11, 103, 118, 134]. In dogs, basal SS levels are influenced by vagal, cholinergic and adrenergic mechanisms and by prostaglandins, whereas histaminergic stimulation of H_2 receptors has no effect (Table 1) [95]. In the interdigestive period, there is a re-

Table 1. *Effects of vagal and sympathetic nerves, cholinergic muscarinic and histaminergic mechanisms and prostaglandins on basal SS levels in dogs*

	Peripheral vein	*Antral vein*	*Fundic vein*	*Pancreatic vein*
Vagal nerves	↓	↓	↑	0
Sympathetic nervous system	↑	↑	0	↑
Cholinergic muscarinic mechanisms	↑	0	↑	↑
Histamine H_2 receptors	0	0	0	0
Prostaglandins	↑	↑	↑	↑

↑, increase in basal SS level; ↓, decrease in basal SS level; 0, no change in basal SS level

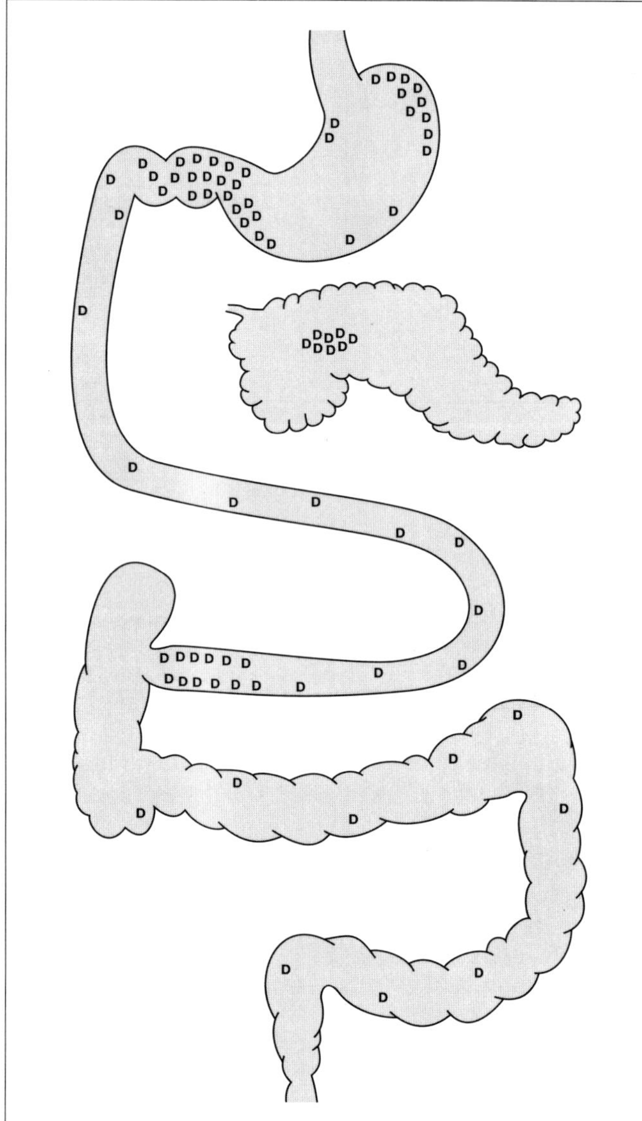

Fig. 1. *Distribution of D cells in the GI tract and pancreas*

lationship between the fluctuations of basal SLI levels and the motor activity in the GI tract that suggests a close association between motilin and SS in the fasted state [3].

Postprandial Somatostatin Release

The ingestion of a mixed meal elicits a significant increase in peripheral vein SS levels, as shown initially in dogs [12, 103, 105, 109, 118] and subsequently in humans [77, 133, 134, 136]. Although gastric and pancreatic SS secretion increases after a meal [103], the contribution of pancreatic D-cell activation to the rise in peripheral plasma SS levels is negligible, as shown in man and dog [31, 126]. It can thus be assumed that this rise is solely of gastric and intestinal origin. Interestingly, it has been shown that in pigs the stomach and pancreas secrete mainly SS-14, whereas the intestine releases mainly SS-28 [7]. The small intestine is the major source of SS-28 present in the peripheral circulation in man in the basal state and after feeding [6, 17, 25, 26, 79, 121]. In response to a meal, both molecular forms increase in the circulation [6, 17, 25, 26, 77, 79, 106, 121], even though their respective contributions remain unclear (Fig. 2).

In rats, an increase of SS levels has been observed in the portal vein [11], but not in the peripheral plasma [96], which might point to a greater importance of paracrine effects in this species compared with the dual endocrine and paracrine effects seen in dog and man.

In dogs and humans, the plasma levels of SS in peripheral veins increase in response to all three basic nutrients – fat, carbohydrate and protein –, but they respond most to protein meals in dogs and to fat-rich meals in humans [103, 118].

Mechanisms of Somatostatin Release

Cephalic Phase

The rise in SS secretion occurs during the cephalic, gastric and intestinal phases of a meal. Even though there is no clear separation between these phases, this classical division will be maintained for the sake of clarity. In dogs, the entire sequence of sham feeding appears to be required for an increase in SS release: the sight and smell of food are insufficient stimuli [19, 95]. In man, modified sham feeding increases SS release [8]. The differential SS response in the fundic and antral areas of the stomach and in the islets of Langerhans is unknown.

During the cephalic phase, the sight, smell and taste of food primarily activate central nervous mechanisms in the hypothalamus and brain stem. This leads to an activation of vagal fibres in the dorsal motor nucleus and the nucleus ambiguus, which in turn activates vagal efferent fibres that innervate the GI tract. In the dog, it can be assumed that vagal stimulation contributes to cephalic-phase SS release, since electrical vagal stimulation activates D-cell function in the GI tract and pancreas [2, 32, 50]. Interestingly, the rise in peripheral SS levels during sham feeding is partly due to cholinergic mechanisms [19], whereas the vagally induced increase in peripheral SS is atropine resistant and appears to be mediated entirely by nicotinic cholinergic receptors [2]. This discrepancy awaits further elucidation.

In other species, the effect of sham feeding on SS release is unknown, but we do have some information about the effect of vagal activation on SS secretion. In man, vagal activation can lead to increased SS levels in the peripheral circulation, since the rise in SS during insulin-induced hypoglycaemia is abolished by truncal vagotomy [29]. In the porcine GI tract, it was shown that vagal stimulation augments antral and inhibits fundic SS secretion [67]. Subsequent studies

Fig. 2. *Following the ingestion of a meal, SS from the fundic and antral areas of the stomach and from the small intestine is released into the circulation, whereas pancreatic SS does not contribute substantially to the postprandial rise in the plasma SS level*

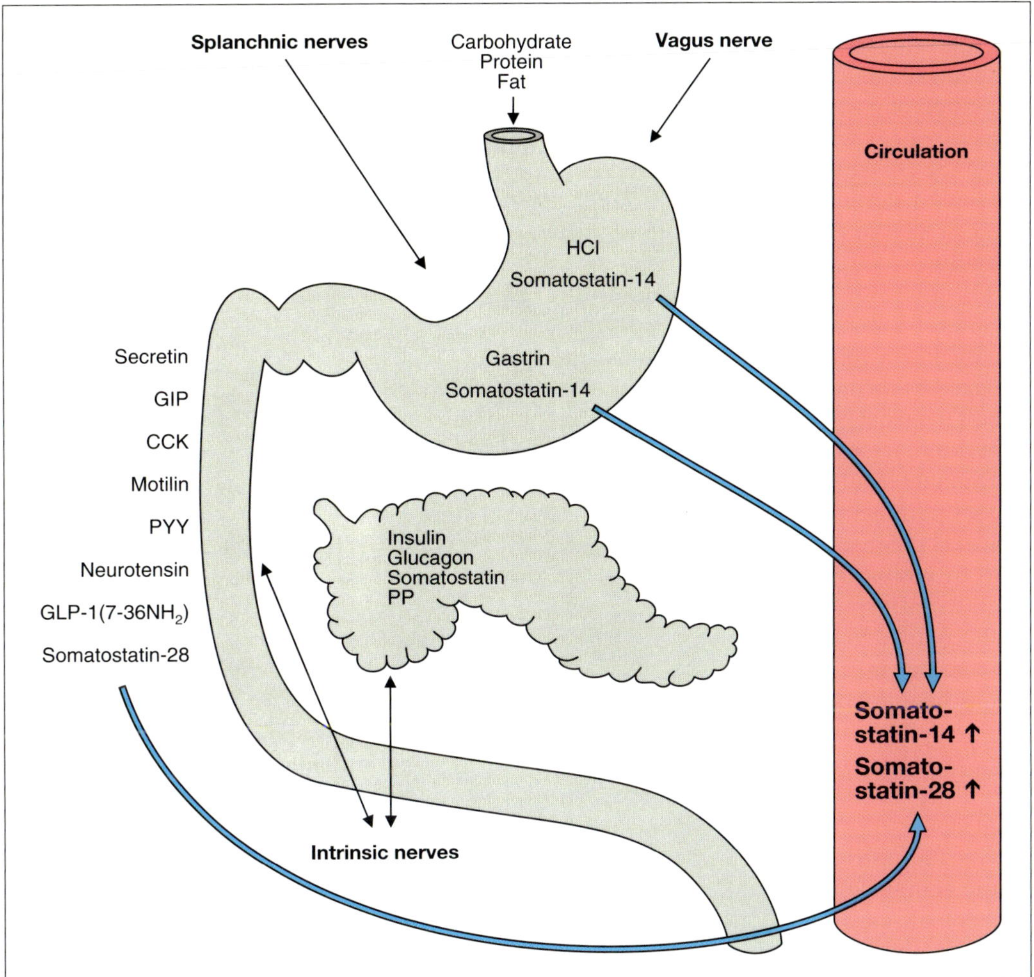

showed that, in this species, both vagally induced SS and gastrin secretion from the antrum are entirely mediated by the neurotransmitter gastrin-releasing peptide (GRP) [39], while cholinergic mechanisms did not prove effective under the conditions of stimulation employed.

By contrast, vagal stimulation in cats and rats inhibits gastric SS release [57, 63, 64, 132]. For technical reasons, differential catheterization of veins draining the antral and fundic regions of the stomach is virtually impossible, and the portal vein measurements performed in these species reflect the activity of both regions. However, studies in rats have shown that antral SS does not contribute to portal vein levels [45, 75].

Since the vagal fibres and the intrinsic neurons of the GI tract contain not only classical neurotransmitters (acetylcholine, noradrenaline, serotonin) but also numerous peptidergic ones, the effects of vagal excitation on endocrine and/or exocrine cells can be mediated through different mechanisms. In the rat stomach, the inhibitory effect of vagal stimulation is an entirely cholinergic muscarinic effect that is very potent and masks any noncholinergic stimulatory effects [57, 63]. In contrast, in the porcine stomach the mammalian bombesin-like peptides GRP and neuromedin C are the major neurotransmitters of vagally induced gastric SS release [38, 39]. Thus, depending on the species examined, classical (cholinergic) or peptidergic neurotransmission may prevail.

Although the cephalic-phase effects are predominantly of vagal origin, a contribution of sympathetic mechanisms cannot be excluded. Splanchnic nerve stimulation has no effect on gastric SS release from the rat and porcine stomach [64, 67].

Gastric Phase

The gastric phase of a meal comprises all the functional changes that are brought about by the presence of food in the stomach.

The major physiological stimuli during the gastric phase are (a) the distension of the stomach by the volume load of the ingested food and (b) the chemical composition of the nutrient mixture. In the basal state, the intragastric pH is 1–2; following food ingestion, it increases to 5–6 as a result of the neutralizing capacity of the meal; the subsequent augmentation of gastric acid secretion then brings the pH back to basal values within 60–120 min [61]. The difference between the late postprandial period and the basal interdigestive period, during both of which the luminal milieu is highly acidic, is that in the former the stomach contains acidified nutrients.

GI distension due to the volume load of the meal does not affect SS secretion in dogs and humans, a fact suggesting that the major role for the postprandial activation of D cells is played by the nutrient content and the chemical composition of the meal [96]. On the other hand, gastric acidity is an important factor for the regulation of GI SS release in dogs and humans [55, 103, 111]. Acidification of a protein meal increases peripheral SS levels substantially during the gastric phase [111]. Since fundic SS is suppressed under these conditions by inhibitory cholinergic mechanisms, this rise is due solely to antral release [103, 111].

In contrast to dog and man, acidification of the intragastric contents is of no importance for gastric SS secretion in rats, either in vivo or in vitro [96, 99].

Neural Mechanisms During the Gastric Phase

During the gastric phase, the ingested foodstuff activates intrinsic neural elements, and under physiological conditions the cephalic-vagal mechanisms reinforce this effect. A similar synergism has been shown for the regulation of gastric acid secretion in humans [80].

In the canine stomach, truncal vagotomy increases the postprandial release of antral SS, while fundic SS release remains unaffected [111]. Studies with atropine have shown that cholinergic mechanisms participate in the stimulation of antral SS release, whereas their role in the regulation of fundic SS release depends on the intragastric pH. When the nutrients have a neutral pH (early postprandial period), cholinergic mechanisms contribute to the stimulation of fundic SS release; when they are acidified (later postprandial period), however, cholinergic mechanisms participate in the inhibition of fundic SLI [111]. As in dogs, atropine reduced the postprandial SS response in man [54], an effect most probably independent of intragastric acidity.

Although adrenergic mechanisms do not appear to be of importance for cephalic-phase SS release, the sympathetic nervous system contributes to the postprandial SS response. In chemically sympathectomized dogs, gastric and pancreatic SS secretion and peripheral SS levels are reduced [110]. By contrast, neither α-adrenergic nor β-adrenergic blocking agents modify the postprandial SS response in humans [54].

Table 2 summarizes the effects of various neuropeptides of the intrinsic nervous system and local tissue factors such as prostaglandins and histamine. It is interesting that only the tachykinins (substance P and neurokinins A and B) have an inhibitory effect, whereas most other peptides stimulate SS release [reviews in 95, 96, 114; also 5, 14, 22, 49, 59, 92, 112, 115]. The effect of vasoactive intestinal peptide (VIP) is dose dependent, i. e. it is inhibitory at lower doses and stimulatory at higher doses [92].

Table 2. *Possible effects of neuropeptides of the intrinsic nervous system and local tissue factors on postprandial gastric SS release*

Stimulation	*Inhibition*
Met-enkephalin	Substance P
Growth-hormone-releasing factor (GHRF)	Neurokinin A
Vasoactive intestinal peptide (VIP)	Neurokinin B
Peptide histidine-isoleucine (PHI)	Vasoactive intestinal peptide (VIP)
Calcitonin-gene-related peptide (CGRP)	
Histamine	
Prostaglandins	

Conclusive evidence of a physiological role of most of these peptides awaits the development of specific peptide-receptor antagonists. The present list is based on infusion studies of these peptides in vitro or in vivo

Intestinal Phase

The intestinal phase of a meal comprises all the functions that are activated by the presence of food in the small bowel. During this phase, the nutrients have to be digested by enzymes released from the exocrine pancreas. Furthermore, the endocrine pancreas is stimulated by the absorbed nutrients and intestinal hormones, and finally intestinal factors exert an inhibitory effect on gastric acid secretion, gastrin secretion and gastric emptying.

Intestinal Phase and Gastric Somatostatin

In dogs, fundic and antral SS release is stimulated by intraduodenally administered nutrients such as glucose, fat and protein, with a parallel increase in peripheral venous SS levels [113]. Intraduodenal acidification also activates gastric SS release [103]. In humans, intestinally administered fat is a potent stimulus of GI SS secretion, protein and carbohydrate being less effective [53]. Since nutrients in the intestine stimulate the release of GI hormones such as secretin, cholecystokinin (CCK), gastric inhibitory peptide (GIP), glucagon-like peptide (7–36) amide (GLP-1, 7–36-NH_2) and peptide YY (PYY), these hormones can be assumed to mediate the intestinally induced gastric SS response. Indeed, they have been shown in vitro and in vivo to be effective stimulators of gastric SS secretion [24, 63, 70, 81, 91].

Neural factors also play a role. Truncal vagotomy elicits an increase in SS release during an intraduodenal protein meal, which suggests that inhibitory vagal fibres are activated by intestinal mechanisms [113]. This inhibitory vagal effect is counterbalanced by stimulatory cholinergic mechanisms [113].

During the intestinal phase of a meal, nutrients such as glucose and amino acids are rapidly absorbed into the peripheral circulation. The postprandial elevation in their levels might thus play a role in the regulation of not only the endocrine pancreas but also the GI tract. As mentioned above, peripheral SS levels reflect GI rather than pancreatic SS secretion.

The increase in plasma SS levels after mixed carbohydrate/fat or carbohydrate/protein meals is less than the response to fat or protein meals alone [104, 108]. Since the elevation of circulating glucose levels by intravenous glucose elicits a similar reduction in the SS response to fat or protein meals [108], it is most likely that absorbed and circulating glucose, but not the glucose present in the GI tract, is responsible for this effect. This view is supported by studies showing that a physiological increase in plasma glucose levels also reduces plasma SS after stimulation with intravenous acetylcholine, CCK or motilin [89, 107, 116]. This suggests that glucose reduces SS levels not so much by altering the GI processing of food (effect on intestinal passage, gastric emptying or absorption) as by interacting directly with the D cells. Interestingly, the elevation of glucose levels in the vascular perfusate of an isolated stomach has no effect on basal SS secretion [114]. Moreover, the elevation of glucose in this model abolishes, or even reverses, the inhibitory effects of vagal stimulation or VIP infusion [58, 92]. Thus, there is no evidence for a direct inhibitory action of glucose on the gastric D cell when the organ is outside its physiological milieu; this, of course, supports the notion that the effect of glucose seen in vivo is mediated by other factors, most likely insulin.

Studies with insulin have shown its potent inhibitory action on both gastric and pancreatic D-cell function in dogs [82]. The peripheral venous SS levels are suppressed by insulin in humans too, provided that euglycaemia is maintained [29, 137]. In addition, dogs with alloxan-induced diabetes have an hypersomatostatinaemia that depends on the degree of insulin deficiency [101, 109] and is related to the catabolic state, since free fatty acids and β-hydroxybutyrate stimulate SS secretion [35, 135].

In this context, the endogenous opioids of the intrinsic nervous system have to be regarded as important mediators of the modulatory effects of glucose and/or insulin on SS release. In the dog and in man, the effect of endogenous opioids is stimulatory in conditions of normoglycaemia, whereas during physiological elevations of plasma glucose and insulin concentrations they restrain postprandial SS release [104, 108].

Whether amino acids are as important modulators of GI neuroendocrine functions as glucose and insulin is not known.

Intestinal Phase and Pancreatic Somatostatin

The pancreatic D cells are most probably not a substantial source of circulating SS. The potential local effects of pancreatic SS on the peri-insular exocrine tissue will be discussed in a later section.

Following the ingestion of a meal, pancreatic SS output increases [101], but by so little that it does not contribute to peripheral SS levels. Stimulation of pancreatic D-cell function is mediated by gut hormones and increasing levels of circulating nutrients [40, 41, 81, 87, 101, 113]: in contrast to gastric D cells, the modulatory role of pancreatic D cells is driven not only by GI but also by metabolic stimuli.

Physiological Effects of Somatostatin

The release of SS from its various sources in the GI tract and the ensuing elevation of plasma SS levels in the peripheral circulation suggest that SS can act as a true endocrine factor; indeed, intravenous infusions of SS that reproduce postprandial levels affect several GI and pancreatic functions [97]. On the other hand, the close proximity between D cells and other exocrine or endocrine cells supports the notion of a paracrine mode of action of SS via diffusion through the interstitial space to its target cells [16]. The paracrine mode of action is of far greater importance in the rat than in species such as dog and man, in which the endocrine pathway too has to be regarded as physiologically relevant.

Somatostatin and Gastric Functions

The ingested foodstuff passes rapidly through the oropharynx and the oesophagus into the stomach, where it remains to be processed until the prerequisites for its release into the duodenum are fulfilled. The major task of the stomach is the grinding of ingested food particles and their adequate liquefaction. The material released from the stomach into the duodenum has a particle size of less than 1 mm. The intragastric processing of food is guaranteed by appropriate fluid and acid secretion and coordinated fundic, antral and pyloric motility [20].

Shortly after the discovery of SS, infusion studies showed that it is a potent inhibitor of virtually all gastric exocrine, endocrine and motor functions [review in 94]. Subsequent studies have demonstrated the physiological importance of SS. In dog and man, low-dose SS infusion reduces gastric acid secretion, a fact indicating that SS is a hormonal regulator of parietal cell function [15, 52, 130]. In the rat stomach, evidence has been obtained that fundic D cells exert a paracrine inhibitory effect on parietal cell function during stimulation with pentagastrin, whereas the basal secretion rate is not reduced by SS [122] (Fig.3).

SS can affect the parietal cell function by various mechanisms: (1) SS can reduce acid secretion by acting directly on the parietal cell, as shown on isolated canine or rat parietal cells [74, 88], and (2) SS can block the major physiological stimuli of parietal cell function. Locally released SS inhibits basal gastrin secretion in the rat stomach [83, 122]. In dogs, the postprandial rise in SS attenuates the concomitant gastrin response [117]. In humans, too, SS is a hormonal regulator of gastrin secretion [15] (Fig.3). SS not only inhibits the release of gastrin stored in secretory granules but also reduces gastrin synthesis by directly inhibiting the expression of the gastrin gene [44]. The release of the stimulators histamine and acetylcholine is inhibited by SS [30, 86]. The release of the bombesin-like peptides, recently shown to be

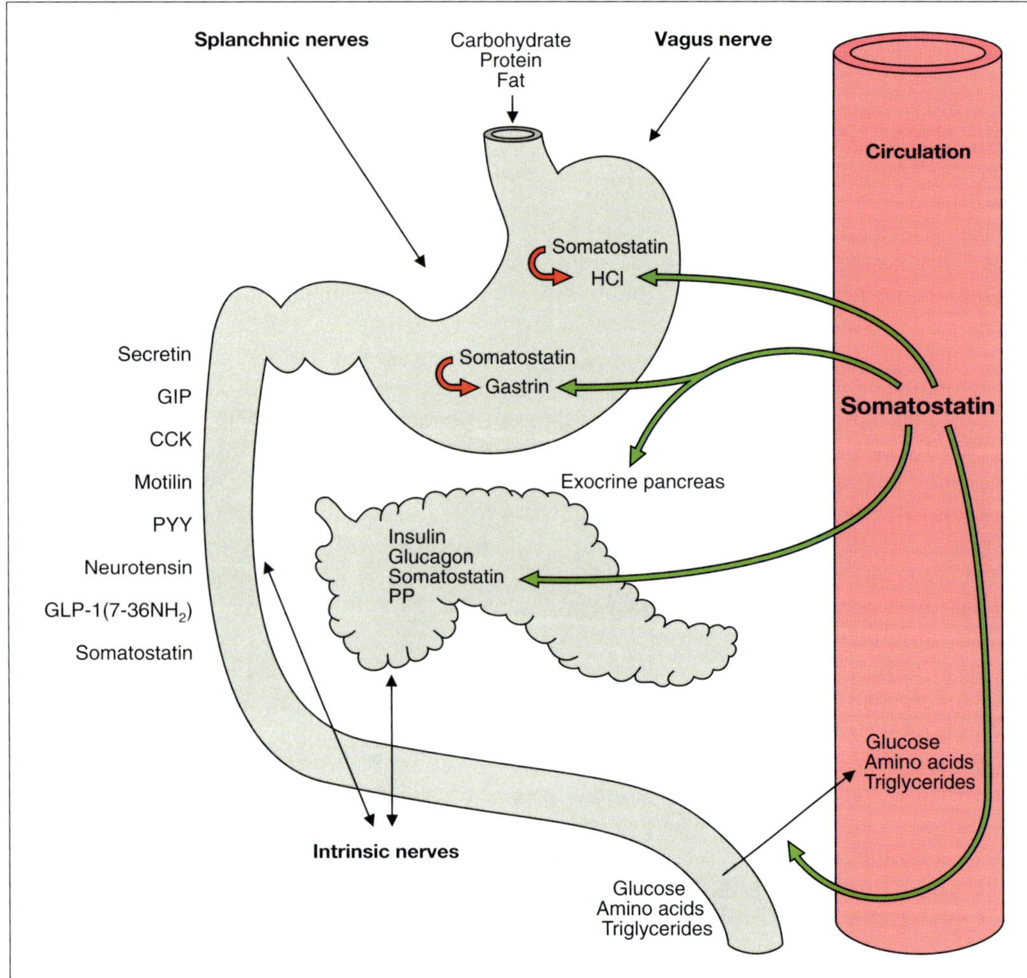

Fig. 3. *In the postprandial phase, SS is released into the circulation and presumably into the interstitial space. Thus, SS might exert not only endocrine* ***(green arrows)*** *but also paracrine* ***(red arrows)*** *control over various exocrine and endocrine functions and over the influx of nutrients from the gut into the bloodstream*

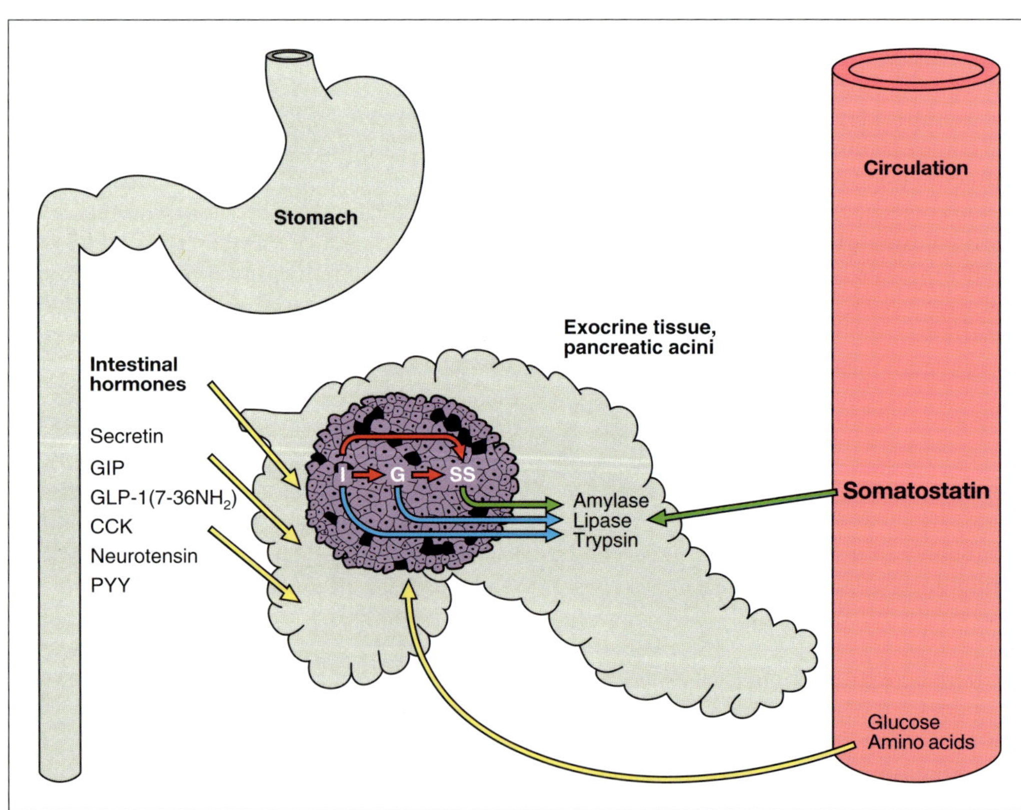

Fig. 4. *During the intestinal phase of a meal, pancreatic exocrine and endocrine function is stimulated by intestinal hormones and the endocrine pancreas also by circulating nutrients (glucose, amino acids). The increase in pancreatic enzyme secretion is controlled by (a) SS from the peripheral circulation and (b) SS and the other islet hormones that are passing through the peri-insular acinar tissue*

important peptidergic stimuli of acid secretion [60], can be reduced by SS [99]. Thus, SS is an important regulator of gastric exocrine and endocrine functions that prevents the generation of large volumes of strongly acidified chyme. If such chyme were released into the duodenum at normal rates of gastric emptying, the neutralizing capacity of the pancreas and small bowel and subsequently the pancreatic digestion of nutrients would be inadequate, with resulting maldigestion of the foodstuff. Such gastric hypersecretion and its consequences are seen in the Zollinger-Ellison syndrome, which is caused by tumour-associated hypergastrinaemia.

As mentioned above, gastric secretion is reduced during the later postprandial phase, which extends from 90 to 120 min after food ingestion. This is due, at least in part, to the acid-induced reduction in gastrin levels that is mediated by increasing levels of SS [103, 111]; other mechanisms, e.g. the inhibitory action of prostaglandins in dogs [112] or histamine in rats [10], play a part as well.

A mechanism that has to be considered in addition to those of gastric origin is that involving "enterogastrone", a term that encompasses all inhibitory effects on parietal cells that originate in the small bowel, especially after intestinal fat administration [47, 129]. At least in some species, several intestinal hormones such as secretin, GIP, GLP-1 and PYY are potential physiological inhibitors of gastric acid secretion [1, 13, 23, 66, 73, 90, 139]. All of them stimulate the gastric secretion of SS, which could thus be the final common mediator of their inhibitory effects. Apart from the local mediator function of SS in the fundic-corpus area of the stomach, it has been suggested that intestinally released SS is a hormonal inhibitor of gastric acid secretion [53, 119, 120]; this hypothesis, however, has recently been challenged [65].

Somatostatin and Pancreatic Functions

The rate of gastric emptying is important not only for the velocity of food processing by digestive enzymes in the small bowel but also for postprandial endocrine functions of the pancreas. This, in turn, has some impact on the regulation of digestive functions and the activity of intracellular key enzymes of intermediary metabolism. Therefore, the slowing of gastric emptying by SS has to be considered a major regulatory function of this hormone [42, 43].

Exocrine Pancreas

The rates of pancreatic enzyme and bicarbonate secretion are important for the intestinal digestion of nutrients; for an optimal action of pancreatic enzymes, they must be properly adjusted to the rate of gastric emptying into the duodenum. The luminal pH of the intestinal contents should be close to neutrality, since the optimum pH for all the enzymes involved is 8.5 (a value that, however, is never reached under physiological conditions). Near-neutrality can be accomplished only if the acidic load leaving the stomach is fairly small, since the buffer capacity of bicarbonate is rather weak at pH levels of 5–7 and cannot be compensated by a larger bicarbonate output owing to the limited maximal secretory capacity of the pancreas.

Another possibility to regulate nutrient assimilation is the reduction of nutrient breakdown by pancreatic digestive enzymes. In dogs and man, the rise of SS in the circulation diminishes pancreatic enzyme and bicarbonate secretion and reduces gallbladder emptying [33, 42]. These mechanisms would be expected to affect predominantly the digestion of ingested fat.

Endocrine Pancreas: Effects on the Exocrine Pancreas

Studies discussed in earlier sections were related to the endocrine role of SS originating in the GI tract and reaching the exocrine pancreas via the systemic circulation. It is necessary, however, to consider SS from the D cells of the endocrine pancreas as another regulatory component, at least for the peri-insular exocrine tissue; evidence for such a potential regulatory loop has been gained from studies on the rat pancreas.

The concept of an insuloacinar axis is based morphologically on the demonstration of an insuloacinar portal blood system and functionally on the appreciation that islet peptides directly regulate acinar cell function. Pancreatic islets contain four major types of endocrine cells: B cells secrete insulin, A cells secrete glucagon, D cells secrete SS, and PP cells secrete pancreatic polypeptide. B cells are predominant by far, accounting for about 80% of islet cells. In the rat and rabbit (but not in all species), B cells occupy the centre of each islet, with the other three cell types scattered about the periphery [68]. The acinar cells in the pancreas are in close contact with the islets, since no significant capsule or basement membrane surrounds the islets. Peri-insular acinar tissue is morphologically different from more remote acinar tissue [34, 48]. There are also differences in the relative concentration of digestive enzymes in peri-insular and other acinar cells [9, 62].

In virtually all mammalian species, the islets of Langerhans have a distinct arterial blood supply. Pancreatic intralobular arteries give a branch to each islet as a vas afferens that divides within the islet into a capillary glomerulus; numerous vasa efferentia emerge from the glomerulus into the surrounding tissue as insuloacinar portal vessels [27, 28, 127, 138]. Because of this pattern of blood flow, the peri-insular exocrine pancreas receives a substantial part of its blood supply from the islet tissue and is therefore exposed to much higher concentrations of islet cell hormones than could be expected to reach the pancreas via the systemic arterial blood supply. The biological relevance of this portal system is supported by studies on the isolated pancreas showing that stimulation of endogenous insulin secretion by glucose selectively increases the secretion of amylase from the exocrine tissue [46, 72, 84]. Accordingly, the possibility exists that the other islet cell hormones, especially SS, also modulate pancreatic exocrine function rather than modify the function of neighbouring islet cells.

Studies in the rat pancreas support this notion. While insulin secretion from B cells has a regulatory influence on the more peripherally located A and D cells, islet SS affects the release of neither insulin nor glucagon owing to the anatomical organization of blood flow outlined above and the downstream position of the D cells in relation to A and B cells [85, 124, 125]. The numerous studies that have shown an alteration of insulin or glucagon secretion by an anti-SS serum in isolated islet preparations have ignored the importance of the vascular microarchitecture and have created an artificial peri-insular medium into which islet cell hormones are released and from which they can diffuse back to all other islet cells.

In view of all these studies, islet SS is most likely a modulator of pancreatic exocrine function. Thus, the peri-insular part of the exocrine pancreas is influenced by higher concentrations of islet SS together with the elevated levels of insulin and/or glucagon, while the more remote part of the exocrine pancreas is modulated by the arterial SS level that is predominantly of GI origin. About 80% of acinar cells receive their blood supply directly from the systemic circulation [51]. This dual regulation makes possible fine tuning of the digestive capacity of the pancreas juice according to the nutrient load in the gut and the metabolic setting determined by the concentrations of circulating nutrients and pancreatic hormones (Fig. 4).

From the studies mentioned above it is evident that the secretion of insulin and glucagon is modified by circulating SS that originates in the GI tract. This is supported by various intravenous infusion studies in dogs and humans in which physiological elevations of SS-14 and SS-28 in the plasma have been shown to reduce the secretion of insulin and glucagon [18, 71, 117, 123, 140].

Somatostatin and Nutrient Entry into the Circulation

The release of SS from its various sources after the ingestion of a meal and the biological effects of exogenous and endogenous SS on numerous exocrine and endocrine functions of the GI tract and pancreas suggest that SS exerts a modulatory effect that ultimately leads to an attenuated and smoothed nutrient influx from the gut into the circulation [128] (Fig. 5). Thus, in dogs acute hyposomatostatinaemia augments the postprandial rise in plasma triglyceride levels [117]. Conversely, physiological elevations of SS by intraportal infusion of synthetic SS reduce the postprandial blood levels of carbohydrate and triglycerides [102].

It is noteworthy that obese subjects have reduced basal plasma levels of SS and virtually no SS response to a meal. This hyposomatostatinaemia is associated with hyperinsulinaemia [106]. Replacement of SS by intravenous infusion rapidly reduces basal and postprandial plasma insulin levels.

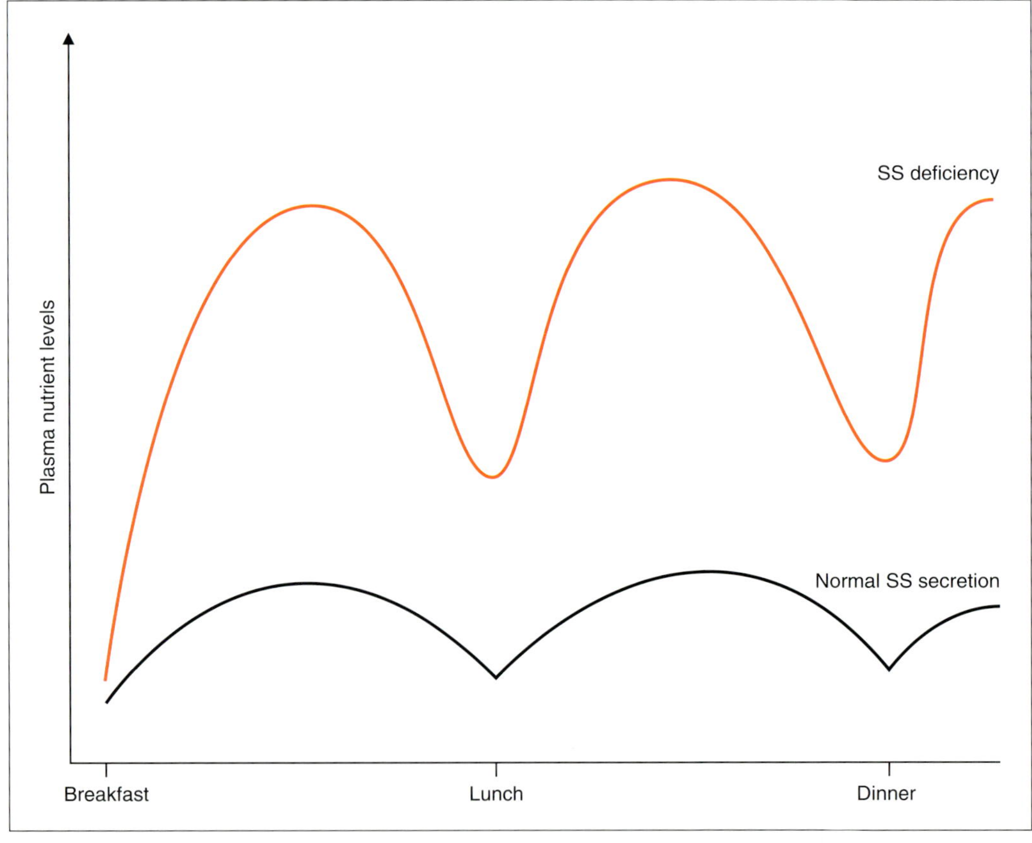

Fig. 5. *The rise in SS after the ingestion of a meal prevents an exaggerated exocrine and endocrine secretory response in the GI tract and retards the rate at which glucose, amino acids and triglycerides enter the circulation. The elevations of nutrient levels in the blood are thereby made smoother*

Since insulin is a potent inhibitor of SS, the hyposomatostatinaemia might very well be secondary to the hyperinsulinaemia, but on the other hand it could cause even higher insulin levels and increased nutrient influx in obese subjects. This vicious circle is favoured by the ingestion of highly refined carbohydrates, which poorly stimulate SS secretion but potently stimulate insulin secretion; in liquid form particularly, they are rapidly released from the stomach and cause a rate-dependent increase in insulin but not SS secretion [100]. Thus, rapidly expanding nutritional habits due to modern techniques of food preparation cannot be counterbalanced by physiological regulation systems for which there was no need until recently, i. e. until 200–300 years ago.

In conclusion, SS located in the stomach, small intestine and pancreas is released after the ingestion of food. Its secretion is tightly integrated in the complex regulatory system of neuroendocrine factors that convey information about the composition and amount of ingested foodstuffs from the gut lumen to the digestive and metabolically relevant organs. This provides the control over all avenues of nutrient flux necessary to ensure the homeostasis of the *milieu intérieur.*

References

1. Adrian TE, Savage AP, Sagor GR, Allen JM, Bacarese-Hamilton AJ, Tatemoto K, Polak JM, Bloom SR (1985) Effect of peptide YY on gastric, pancreatic, and biliary function in humans. Gastroenterology 89: 494–499
2. Ahrén B, Paquette L, Taborsky GJ Jr (1986) Effect and mechanism of vagal nerve stimulation on somatostatin secretion in dogs. Am J Physiol 250: E212–E217
3. Aizawa I, Itoh Z, Harris V, Unger RH (1981) Plasma somatostatin-like immunoreactivity during the interdigestive period in the dog. J Clin Invest 68: 206–213
4. Alumets J, Sundler F, Hakanson R (1977) Distribution, ontogeny and ultrastructure of somatostatin immunoreactive cells in the pancreas and gut. Cell Tissue Res 185: 465–469
5. Amatani T, Kadowaki S, Chiba T, Abe H, Chihara K, Fukase M, Fujita T (1986) Calcitonin gene-related peptide stimulates somatostatin release from isolated perfused rat stomach. Endocrinology 118: 2144–2147
6. Baldissera FGA, Munoz-Perez MA, Holst JJ (1983) Somatostatin 1–28 circulates in human plasma. Regul Pept 6: 63–69
7. Baldissera FGA, Nielsen OV, Holst JJ (1985) The intestinal mucosa preferentially releases somatostatin-28 in pigs. Regul Pept 11: 251–262
8. Befrits R, Uvnäs-Moberg K, Johansson C (1990) Interactions between antral peptides and prostaglandin biosynthesis in gastric acid regulation in man. Digestion 45: 9–18
9. Bendayan M, Ito S (1979) Immunohistochemical localization of exocrine enzymes in normal rat pancreas. J Histochem Cytochem 277: 1029–1034
10. Bender H, Pfeffer A, Schusdziarra V (1987) Effect of indomethacin on bombesin-like immunoreactivity, somatostatin and gastrin secretion from rat stomach. Clin Physiol Biochem 5: 268–275
11. Berelowitz M, Kronheim S, Pimstone B, Shapiro B (1978) Somatostatin-like immunoreactivity in rat blood. J Clin Invest 61: 1410–1414
12. Chayvialle JA, Miyata M, Rayford PL, Thompson JC (1980) Effects of test meal, intragastric nutrients, and intraduodenal bile on plasma concentrations of immunoreactive somatostatin and vasoactive intestinal peptide in dogs. Gastroenterology 79: 844–852
13. Chey WY, Kim MS, Lee KY, Chang TM (1981) Secretin is an enterogastrone in the dog. Am J Physiol 240: G239–G244
14. Chiba T, Taminato T, Kadowaki S, Inoue Y, Mori K, Seino Y, Abe H, Chihara K, Matsukura S, Fujita T, Goto Y (1980) Effects of various gastrointestinal peptides on gastric somatostatin release. Endocrinology 106: 145–149
15. Colturi TJ, Unger RH, Peters M, Feldman M (1983) Physiologic role for circulating somatostatin in gastric secretion in man. Gastroenterology 84: 60–65
16. Creutzfeldt W (1987) Evidence for paracrine function of somatostatin. In: Reichlin S (ed) Somatostatin: basic and clinical status. Plenum, New York, pp 201–208
17. D'Alessio DA, Ensinck JW (1990) Fasting and postprandial concentrations of somatostatin-28 and somatostatin-14 in type II diabetes in men. Diabetes 39: 1198–1202
18. D'Alessio DA, Sieber C, Beglinger C, Ensinck JW (1989) A physiologic role for somatostatin 28 as a regulator of insulin secretion. J Clin Invest 84: 857–862
19. De Graef J, Woussen-Colle MC (1985) Effects of sham feeding, bethanechol, and bombesin on somatostatin release in dogs. Am J Physiol 248: G1–G7
20. Dent J (1990) Patterns of coordination of pyloric contractions with those of the antrum and duodenum. In: van Nueten JM, Schuurkes JAJ, Akkermans LMA (eds) Gastro-pyloro-duodenal coordination. Wrightson, Petersfield, pp 127–138
21. Dubois MP (1975) Immunoreactive somatostatin is present in discrete cells of the endocrine pancreas. Proc Natl Acad Sci USA 72: 1340–1343
22. Dunning BE, Taborsky GJ Jr (1987) Calcitonin gene-related peptide: a potent and selective stimulator of gastrointestinal somatostatin secretion. Endocrinology 120: 1774–1781
23. Ebert R, Aschenbeck J, Creutzfeldt W (1990) Effect of human gastric inhibitory polypeptide (GIP) on gastric emptying and acid secretion. Digestion [Suppl 1] 46: 26
24. Eissele R, Koop H, Arnold R (1990) Effect of glucagon-like peptide-1 on gastric somatostatin and gastrin secretion in the rat. Scand J Gastroenterol 25: 449–454
25. Ensinck JW, Laschansky EC, Vogel RE, Simonowitz DA, Roos BA, Francis BH (1989) Circulating prosomatostatin-derived peptides. J Clin Invest 83: 1580–1589
26. Ensinck JW, Vogel RE, Laschansky EC, Francis BH (1990) Effect of ingested carbohydrate, fat, and protein on release of somatostatin-28 in humans. Gastroenterology 98: 633–638
27. Fujita T (1973) Insulo-acinar portal system in the horse pancreas. Arch Histol Jpn 35: 161–171
28. Fujita T, Murakami T (1973) Microcirculation of monkey pancreas with special reference to the insulo-acinar portal system. A scanning electron microscope study of vascular casts. Arch Histol Jpn 35: 255–263
29. Glaser B, Vinik AI, Valtysson G, Hoghlin G (1981) Truncal vagotomy abolishes the somatostatin response to insulin-induced hypoglycemia in man. J Clin Endocrinol Metab 52: 823–825
30. Guillemin R (1976) Somatostatin inhibits the release of acetylcholine induced electrically in the myenteric plexus. Endocrinology 99: 1653–1654
31. Gutniak M, Grill V, Wiechel KL (1987) Basal and meal-induced somatostatin-like immunoreactivity in healthy subjects and in IDDM and totally pancreatectomized patients. Diabetes 36: 802–808
32. Guzman S, Chayvialle JA, Banks WA, Rayford PL, Thompson JC (1979) Effect of vagal stimulation on pancreatic secretion and blood levels of gastrin, cholecystokinin, secretin, vasoactive intestinal peptide and somatostatin. Surgery 86: 329–335
33. Gyr K, Beglinger C, Köhler E, Trautzl U, Keller U, Bloom SR (1987) Circulating somatostatin – physiological regulator of pancreatic function? J Clin Invest 79: 1595–1600

34. Hellman B, Wallgren A, Petersson B (1962) Cytological characteristics of the exocrine pancreatic cells with regard to their position in relation to the islets of Langerhans. Acta Endocrinol (Copenh) 93: 465–473
35. Hermansen K (1982) Stimulatory effect of β-hydroxybutyrate on the release of somatostatin from the isolated pancreas of normal and streptozotocin-diabetic dogs. Diabetes 31: 270–274
36. Hökfelt T, Efendic S, Hellerström C, Johansson O, Luft R, Arimura A (1975) Cellular localization of somatostatin in endocrine-like cells and neurons of the rat with special references to the A-cells of the pancreatic islets and to the hypothalamus. Acta Endocrinol [Suppl 200] (Copenh) 80: 5–41
37. Hökfelt T, Elfvin LG, Elde E, Schultzberg M, Goldstein M, Luft R (1977) Occurrence of somatostatin-like immunoreactivity in some peripheral sympathetic noradrenergic neurons. Proc Natl Acad Sci USA 774: 3587–3591
38. Holst JJ, Jensen SL, Knuhtsen S, Nielsen OV, Rehfeld JF (1983) Effect of vagus, gastric inhibitory polypeptide, and HCl on gastrin and somatostatin release from perfused pig antrum. Am J Physiol 244: G515–G522
39. Holst JJ, Knuhtsen S, Orskov C, Skak-Nielsen T, Poulsen SS, Nielsen VV (1987) GRP-producing nerves control antral somatostatin and gastrin secretion in pigs. Am J Physiol 253: G767–G774
40. Ipp E, Dobbs RE, Arimura A, Vale W, Harris V, Unger RH (1977) Release of immunoreactive somatostatin from the pancreas in response to glucose, amino acids, pancreozymin-cholecystokinin and tolbutamide. J Clin Invest 60: 760–765
41. Ipp E, Dobbs RE, Harris V, Arimura A, Vale W, Unger RH (1977) The effects of gastrin, gastric inhibitory polypeptide, secretin and the octapeptide of cholecystokinin upon immunoreactive somatostatin release by the perfused canine pancreas. J Clin Invest 60: 1216–1219
42. Johansson C, Kollberg B, Efendic S, Uvnas-Wallensten K (1981) Effects of graded doses of somatostatin on gallbladder emptying and pancreatic enzyme output after oral glucose in man. Digestion 22: 24–31
43. Johansson C, Wisen O, Efendic S, Uvnas-Wallensten K (1981) Effect of somatostatin on gastrointestinal propagation and absorption of oral glucose in man. Digestion 22: 126–137
44. Karnik PS, Monahan SJ, Wolfe MM (1989) Inhibition of gastrin gene expression by somatostatin. J Clin Invest 83: 367–372
45. Koop H, Behrens I, Bothe E, McIntosh CHS, Pederson RA, Arnold R, Creutzfeld W (1982) Adrenergic and cholinergic interactions in rat gastric somatostatin and gastrin release. Digestion 25: 96–102
46. Korc M, Owerbach D, Quinto C, Rutter WJ (1981) Pancreatic islet-acinar cell interaction: amylase messenger RNA levels are determined by insulin. Science 213: 352–353
47. Kosaka T, Lim RKS (1930) Demonstration of the humoral agent in fat inhibition of gastric secretion. Proc Soc Exp Biol Med 27: 890–891
48. Kramer MF, Tan HT (1968) The peri-insular acini of the pancreas of the rat. Z Zellforsch 86: 163–170
49. Kwok YN, McIntosh CHS, Sy H, Brown JC (1988) Inhibitory actions of tachykinins and neurokinins on release of somatostatin-like immunoreactivity from the isolated perfused rat stomach. J Pharmacol Exp Ther 246: 726–731
50. Lefebvre PJ, Luyckx AS (1980) Extrapancreatic glucagon: experimental studies using the isolated perfused dog stomach. In: Andreani D, Lefebvre PJ, Marks V (eds) Current views on hypoglycemia and glucagon. Academic, London, pp 27–35
51. Lifson N, Kramlinger KG, Mayrand RR, Lender EJ (1980) Blood flow to the rabbit pancreas with special reference to the islet of Langerhans. Gastroenterology 79: 466–473
52. Loud FB, Holst JJ, Egense E, Petersen B, Christiansen J (1985) Is somatostatin a humoral regulator of the endocrine pancreas and gastric acid secretion in man? Gut 26: 445–449
53. Lucey MR, Fairclough PD, Wass JAH, Kwasowski P, Medbak S, Webb J, Rees LH (1984) Response of circulating somatostatin, insulin, gastrin and GIP to intraduodenal infusion of nutrients in normal man. Clin Endocrinol (Oxf) 21: 209–217
54. Lucey MR, Wass JAH, Fairclough P, Webb J, Webb S, Medbak S, Rees LH (1985) Autonomic regulation of postprandial plasma somatostatin, gastrin, and insulin. Gut 26: 683–688
55. Lucey MR, Wass JAH, Rees LH, Dawson AM, Fairclough PD (1989) Relationship between gastric acid and elevated plasma somatostatin-like immunoreactivity after a mixed meal. Gastroenterology 97: 867–872
56. Luft R, Efendic S, Hokfelt T, Johansson O, Arimura A (1974) Immunohistochemical evidence for the localization of somatostatin-like immunoreactivity in a cell population of the pancreatic islets. Med Biol 52: 428–430
57. Madaus S, Bender H, Schusdziarra V, Kehe K, Munzert G, Weber G, Classen M (1990) Vagally induced release of gastrin, somatostatin and bombesin-like immunoreactivity from perfused rat stomach. Effect of stimulation frequency and cholinergic mechanisms. Regul Pept 30: 179–192
58. Madaus S, Schusdziarra V, Dummer W, Classen M (1991) The effect of glucose and insulin on vagally induced gastrin, bombesin-like immunoreactivity and somatostatin secretion from the perfused rat stomach. Neuropeptides 18: 215–222
59. Madaus S, Schusdziarra V, Seufferlein T, Classen M (1991) Comparison of α- and β-CGRP on somatostatin and gastrin secretion in the isolated rat stomach. Eur J Gastroenterol Hepatol 3: 35–40
60. Mailliard ME, Wolfe MM (1989) Effect of antibodies to the neuropeptide GRP on distention-induced gastric acid secretion in the rat. Regul Pept 26: 287–296
61. Malagelada JR, Longstreth GF, Summerskill WHJ, Go VLW (1976) Measurement of gastric functions during digestion of ordinary solid meals in man. Gastroenterology 70: 203–210
62. Malaisse-Lagae F, Ravazzola M, Robberecht P, Vandermeers A, Malaisse WJ, Orci L (1975) Exocrine pancreas: evidence for topographic partition of secretory function. Science 190: 795–797
63. McIntosh C, Pederson RA, Koop H, Brown JC (1981) Gastric inhibitory polypeptide stimulated secretion of somatostatin-like immunoreactivity from the stomach, inhibition by acetylcholine or vagal stimulation. Can J Physiol Pharmacol 59: 468–472
64. McIntosh C, Pederson R, Muller M, Brown J (1981) Autonomic nervous control of the gastric somatostatin secretion from the perfused rat stomach. Life Sci 29: 1477–1483
65. Mogard MH, Maxwell V, Wong H, Reedy TJ, Sytnik B, Walsh JH (1988) Somatostatin may not be a hormonal messenger of fat-induced inhibition of gastric functions. Gastroenterology 94: 405–408
66. O'Halloran DJ, Nikou GC, Kreymann B, Ghatei MA, Bloom SR (1990) Glucagon-like peptide-1 (7–36)-NH_2: a physiological inhibitor of gastric acid secretion in man. J Endocrinol 126: 169–173
67. Olesen M, Holst JJ, Sottimano C, Nielsen OV (1987) Autonomic nervous control of fundic secretion of somatostatin and antral secretion of gastrin and somatostatin in pigs. Digestion 36: 24–35
68. Orci L (1982) Macro- and micro-domains in the endocrine pancreas. Diabetes 31: 538–565
69. Orci L, Baetens D, Dubois MP, Rufener C (1975) Evidence for the D-cell of the pancreas secreting somatostatin. Horm Metab Res 7: 400–402
70. Orskov C, Holst JJ, Nielsen OV (1988) Effect of truncated glucagon-like peptide-1 (proglucagon-(78–107amide) on endocrine secretion from pig pancreas, antrum, and nonantral stomach. Endocrinology 123: 2009–2013
71. O'Shaughnessy DJ, Long RG, Adrian TE (1985) Somatostatin-14 modulates postprandial glucose levels and release of gastrointestinal and pancreatic hormones. Digestion 31: 234–242
72. Otsuki M, Williams JA (1982) Effect of diabetes mellitus on the regulation of enzyme secretion by isolated rat pancreatic acini. J Clin Invest 70: 148–156
73. Pappas TN, Debas HT, Goto Y, Taylor IL (1985) Peptide YY inhibits meal-stimulated pancreatic and gastric secretion. Am J Physiol 248: G118–G123
74. Park J, Chiba T, Yamada T (1987) Mechanisms for direct inhibition of canine gastric parietal cells by somatostatin. J Biol Chem 262: 14 190–14 196
75. Pederson RA, McIntosh CHS, Müller MK, Brown JC (1981) Absence of a relationship between immunoreactive-gastrin and so-

matostatin-like immunoreactivity secretion in the perfused rat stomach. Regul Pept 2: 53–60
76. Pelletier G, Leclerc R, Arimura A, Schally AV (1975) Immunohistochemical localization of somatostatin in the rat pancreas. J Histochem Cytochem 23: 699–701
77. Penman E, Wass JAH, Medbak S, Morgan L, Lewis JM, Besser GM, Rees LH (1981) Response of circulating immunoreactive somatostatin to nutritional stimuli in normal subjects. Gastroenterology 81: 692–699
78. Polak JM, Grimelius L, Pearse AGE, Bloom SR, Arimura A (1975) Growth hormone release inhibiting hormone in gastrointestinal and pancreatic D-cells. Lancet 1: 1220–1221
79. Polonsky KS, Shoelson SE, Docherty HM (1983) Plasma somatostatin 28 increases in response to feeding in man. J Clin Invest 71: 1514–1518
80. Richardson CT, Walsh JH, Cooper RA, Feldman M, Fordtran JS (1977) Studies on the role of cephalic-vagal stimulation in the acid secretory response to eating in normal human subjects. J Clin Invest 60: 435–441
81. Rouiller D, Schusdziarra V, Harris V, Unger RH (1980) Release of pancreatic and gastric somatostatin-like immunoreactivity in response to the octapeptide of cholecystokinin, secretin, gastric inhibitory polypeptide and gastrin-17 in dogs. Endocrinology 107: 524–529
82. Rouiller D, Schusdziarra V, Unger RH (1981) Insulin inhibits somatostatin-like immunoreactivity release stimulated by intragastric HCl. Diabetes 30: 735–738
83. Saffouri B, Weir G, Bitar K, Makhlouf G (1979) Stimulation of gastrin secretion from the perfused rat stomach by somatostatin antiserum. Life Sci 25: 1749–1754
84. Saito A, Williams JA, Kanno T (1980) Potentiation of cholecystokinin induced exocrine secretion by both exogenous and endogenous insulin in isolated and perfused rat pancreata. J Clin Invest 65: 777–782
85. Samols E, Stagner JI, Ewart RBL, Marks V (1988) The order of islet microvascular cellular perfusion is B→A→D in the perfused rat pancreas. J Clin Invest 82: 350–353
86. Sandvik AK, Waldum HL (1988) The effect of somatostatin on baseline and stimulated acid secretion and vascular histamine release from the totally isolated vascularly perfused rat stomach. Regul Pept 20: 233–239
87. Schauder P, McIntosh C, Arends G, Arnold R, Frerichs H, Creutzfeldt W (1976) Somatostatin and insulin release from isolated rat pancreatic islets stimulated by glucose. FEBS Lett 68: 225–227
88. Schepp W, Heim KH, Ruoff HJ (1983) Comparison of the effect of PGE_2 and somatostatin on histamine stimulated ^{14}C-aminopyrine uptake and cyclic AMP formation in isolated rat gastric mucosal cells. Agents Actions 13: 200–206
89. Schick R, Schusdziarra V (1985) Modulation of motilin-induced somatostatin release in dogs by naloxone. Peptides 6: 861–864
90. Schjoldager BTG, Mortensen PE, Christiansen J, Orskov C, Holst JJ (1989) GLP-1 (glucagon-like peptide 1) and truncated GLP-1, fragments of human proglucagon, inhibit gastric acid secretion in humans. Dig Dis Sci 34: 703–708
91. Schmid R, Schusdziarra V, Aulehner R, Weigert N, Classen M (1990) Comparison of GLP-1 (7–36amide) and GIP on release of somatostatin-like immunoreactivity and insulin from the isolated rat pancreas. Z Gastroenterol 28: 280–284
92. Schmid R, Schusdziarra V, Classen M (1988) Modulatory effect of glucose on VIP-induced gastric somatostatin release. Am J Physiol 254: E756–E759
93. Schultzberg M, Dreyfus CF, Gershon MD, Hökfelt T, Elde RP, Nilsson G, Said S, Goldstein M (1978) VIP, enkephalin, substance P, and somatostatin-like immunoreactivity in neurons intrinsic to the intestine: immunohistochemical evidence from organotypic tissue cultures. Brain Res 105: 239–244
94. Schusdziarra V (1980) Somatostatin – a regulatory modulator connecting nutrient entry and metabolism. Horm Metab Res 12: 563–577
95. Schusdziarra V (1983) Somatostatin – physiological and pathophysiological aspects. Scand J Gastroenterol [Suppl 82] 18: 69–84
96. Schusdziarra V (1985) Role of somatostatin in nutrient regulation. Adv Exp Biol Med 188: 425–445
97. Schusdziarra V (1987) Evidence for the endocrine role of somatostatin. In: Reichlin S (ed) Somatostatin: basic and clinical status. Plenum, New York, pp 209–217
98. Schusdziarra V (1988) Role of the islets of Langerhans in nutrient entry and metabolism. Nutrition 4: 158–160
99. Schusdziarra V, Bender H, Pfeffer A, Pfeiffer EF (1984) Modulation of acetylcholine-induced secretion of gastric bombesin-like immunoreactivity by cholinergic and histamine H_2-receptors, somatostatin and intragastric pH. Regul Pept 8: 189–198
100. Schusdziarra V, Dangel G, Henrichs I, Klier M, Pfeiffer EF (1982) Rate dependent stimulation of insulin and glucagon but not gastrin and somatostatin release during the intestinal phase of a meal. Horm Metab Res 14: 169–171
101. Schusdziarra V, Dobbs RE, Harris V, Unger RH (1977) Immunoreactive somatostatin levels in plasma of normal and alloxan diabetic dogs. FEBS Lett 81: 69–72
102. Schusdziarra V, Harris V, Arimura A, Unger RH (1979) Evidence for a role of splanchnic somatostatin in the homeostasis of ingested nutrients. Endocrinology 104: 1705–1708
103. Schusdziarra V, Harris V, Conlon JM, Arimura A, Unger RH (1978) Pancreatic and gastric somatostatin release in response to intragastric and intraduodenal nutrients and HCl in the dog. J Clin Invest 62: 509–518
104. Schusdziarra V, Holland A, Maier V, Pfeiffer EF (1984) Effect of naloxone on pancreatic and gastric endocrine function in response to carbohydrate and fat-rich test meals. Peptides 5: 65–71
105. Schusdziarra V, Ipp E, Harris V, Dobbs RE, Raskin P, Orci L, Unger RH (1978) Studies of the physiology and pathophysiology of the pancreatic D-cell. Metabolism [Suppl 1] 27: 1227–1232
106. Schusdziarra V, Lawecki J, Ditschuneit HH, Lukas B, Maier V, Pfeiffer EF (1985) Effect of low-dose somatostatin infusion on pancreatic and gastric endocrine function in lean and obese nondiabetic humans. Diabetes 34: 595–601
107. Schusdziarra V, Lenz N, Schick R, Maier V (1986) Modulatory effect of glucose, amino acids, and secretin on CCK-8-induced somatostatin and pancreatic polypeptide release in dogs. Diabetes 35: 523–529
108. Schusdziarra V, Rewes B, Lenz N, Maier V, Pfeiffer EF (1983) Carbohydrates modulate opiate receptor mediated mechanisms during postprandial endocrine function. Regul Pept 7: 243–254
109. Schusdziarra V, Rouiller D, Harris V, Conlon JM, Unger RH (1978) The response of plasma somatostatin-like immunoreactivity to nutrients in normal and alloxan diabetic dogs. Endocrinology 103: 2264–2273
110. Schusdziarra V, Rouiller D, Harris V, Dey R, Unger RH (1980) Plasma somatostatin-like immunoreactivity in chemically sympathectomized dogs. Horm Metab Res 12: 656–660
111. Schusdziarra V, Rouiller D, Harris V, Unger RH (1979) Gastric and pancreatic release of somatostatin-like immunoreactivity during the gastric phase of a meal. Effects of truncal vagotomy and atropine in the anesthetized dog. Diabetes 28: 658–663
112. Schusdziarra V, Rouiller D, Jaffe BM, Harris V, Unger RH (1980) Effect of exogenous and endogenous prostaglandin E upon gastric endocrine function in dogs. Endocrinology 106: 1620–1627
113. Schusdziarra V, Rouiller D, Pietri A, Harris V, Zyznar E, Conlon JM, Unger RH (1979) Pancreatic and gastric release of somatostatin-like immunoreactivity during the intestinal phase of a meal. Am J Physiol 273: 555–560
114. Schusdziarra V, Schmid R (1986) Physiological and pathophysiological aspects of somatostatin. Scand J Gastroenterol [Suppl 119] 21: 29–41
115. Schusdziarra V, Schmid R, Bender H, Schusdziarra M, Rivier J, Vale W, Classen M (1986) Effect of vasoactive intestinal peptide, peptide histidine isoleucine and growth hormone-releasing factor-40 on bombesin-like immunoreactivity, somatostatin and gastrin release from the perfused rat stomach. Peptides 7: 127–133
116. Schusdziarra V, Stapelfeldt W, Klier M, Maier V, Pfeiffer EF (1981) Effect of physiological increments of blood glucose on plasma somatostatin and pancreatic polypeptide levels in dogs. Regul Pept 2: 211–218

117. Schusdziarra V, Zyznar E, Rouiller D, Boden G, Brown JC, Arimura A, Unger RH (1980) Splanchnic somatostatin: A hormonal regulator of nutrient homeostasis. Science 207: 530–532
118. Schusdziarra V, Zyznar E, Rouiller D, Harris V, Unger RH (1980) Free somatostatin in the circulation: amounts and molecular sizes of somatostatin-like immunoreactivity in portal, aortic and vena caval plasma of fasting and meal-stimulated dogs. Endocrinology 107: 1572–1576
119. Seal AM, Liu E, Buchan A, Brown J (1988) Immunoneutralization of somatostatin and neurotensin: effect on gastric acid secretion. Am J Physiol 255: G40–G45
120. Seal AM, Meloche RM, Liu YQE, Buchan AMJ, Brown JC (1987) Effects of monoclonal antibodies to somatostatin on somatostatin-induced and intestinal fat-induced inhibition of gastric acid secretion in the rat. Gastroenterology 92: 1187–1192
121. Shoelson SE, Polonsky KS, Nakabayashi T, Jaspan JB, Tager HS (1986) Circulating forms of somatostatin-like immunoreactivity in human plasma. Am J Physiol 250: E428–E434
122. Short GM, Doyle JW, Wolfe MM (1985) Effect of antibodies to somatostatin on acid secretion and gastrin release by the isolated perfused rat stomach. Gastroenterology 88: 984–988
123. Souquet JC, Rambliere R, Riou JP, Beylot M, Cohen R, Mornex R, Chayvialle JA (1983) Hormonal and metabolic effects or near physiological increase of plasma immunoreactive somatostatin-14. J Clin Endocrinol Metab 56: 1076–1079
124. Stagner JI, Samols E, Bonner-Weir S (1988) $\beta \rightarrow \alpha \rightarrow \delta$ pancreatic islet cellular perfusion in dogs. Diabetes 37: 1715–1721
125. Stagner JI, Samols E, Marks V (1989) The anterograde and retrograde infusion of glucagon antibodies suggests that A cells are vascularly perfused before D cells within the rat islet. Diabetologia 32: 203–206
126. Taborsky GJ Jr, Ensinck JW (1984) Contribution of the pancreas to circulating somatostatin-like immunoreactivity in the normal dog. J Clin Invest 73: 216–220
127. Thiel A (1954) Untersuchungen über das Gefäß-System des Pankreasläppchens bei verschiedenen Säugern mit besonderer Berücksichtigung des Kapillarknäuels der Langerhansschen Insel. Z Zellforsch 39: 339–372
128. Unger RH, Ipp E, Schusdziarra V, Orci L (1977) Hypothesis: physiologic role of pancreatic somatostatin and the contribution of D-cell disorder to diabetes mellitus. Life Sci 20: 2081–2084
129. Uvnäs B (1971) Role of duodenum in inhibition of gastric acid secretion. Scand J Gastroenterol 6: 113–125
130. Uvnas-Wallensten K, Efendic S, Johansson C, Sjodin L, Cranwell PD (1981) Effect of intra-antral and intrabulbar pH on somatostatin-like immunoreactivity in peripheral venous blood of conscious dogs. The possible function of somatostatin as an inhibitory hormone of gastric acid secretion and its possible identity with bulbogastrone and antral chalone. Acta Physiol Scand 111: 397–408
131. Uvnas-Wallensten K, Efendic S, Luft R (1978) Occurence of somatostatin-like immunoreactivity in the vagal nerves. Acta Physiol Scand 102: 907–908
132. Uvnas-Wallensten K, Efendic S, Roovete A, Johansson C (1980) Decreased release of somatostatin into the portal vein following electrical vagal stimulation in the cat. Acta Physiol Scand 109: 393–398
133. Vinik AJ, Levitt NS, Pimstone B, Wagner L (1981) Peripheral plasma somatostatin-like immunoreactivity responses to insulin hypoglycemia and a mixed meal in healthy subjects and in non-insulin-dependent maturity-onset diabetics. J Clin Endocrinol Metab 52: 330–335
134. Vinik AJ, Shapiro B, Glaser B, Wagner L (1981) Circulating somatostatin in primates. In: Bloom SR, Polak JM (eds) Gut hormones. Livingstone, Edinburgh, pp 371–375
135. Wasada T, Howard B, Dobbs R, Unger RH (1980) Evidence for a role of free fatty acids in the regulation of somatostatin secretion in normal and alloxan diabetic dogs. J Clin Invest 66: 511–516
136. Wass JAH, Penman E, Dryburgh JR, Tsiolakis D, Goldberg PL, Dawson AM, Besser GM, Rees LH (1980) Circulating somatostatin after food and glucose in man. Clin Endocrinol (Oxf) 12: 569–574
137. Webb S, Levy I, Wass JAH, Lorens A, Penman E, Casamitjama R, Wu P, Gaya J, Martinez MJ, Rivera F (1983) Insulin induced somatostatin secretion in man is mediated by gastric acid. Regul Pept [Suppl] 2: S11
138. Wharton GK (1932) The blood supply of the pancreas, with special reference to that of the islets of Langerhans. Anat Rec 53: 55–81
139. Wolfe MM, Hocking MP, Maico DG, McGuigan JE (1983) Effects of antibodies to gastric inhibitory peptide on gastric acid secretion and gastrin release in the dog. Gastroenterology 84: 941–948
140. Zyznar E, Pietri A, Harris V, Unger RH (1981) Evidence for the hormonal status of somatostatin in man. Diabetes 30: 883–886

Therapeutic Use of Somatostatin and Octreotide Acetate in Neuroendocrine Tumors

P. N. Maton[1] and R. F. Arakaki[2]

[1] *Oklahoma Foundation for Digestive Research, Oklahoma City, OK, USA*
[2] *Department of Medicine, University of Hawaii Medical School, Honolulu, Hawaii, USA*

Somatostatin (SS) is a cyclic polypeptide that is produced in different forms and has inhibitory effects on a number of functions. A single gene encodes a SS precursor that is processed to yield two smaller active forms, SS-14 and SS-28 [89]. Both polypeptide forms share the same 14 amino acids at the carboxy terminal, and they appear to be equally potent in producing a biological response [99]. The four amino acids Phe-Trp-Lys-Thr at positions 7–10 in the β turn of the molecule are required for bioactivity [97]. Both the linear (reduced) and the disulfide-linked cyclic forms of SS are biologically active; however, the lesser potency of noncyclic peptides suggests that a specific conformational structure is required for maximal activity. SS-14 and SS-28 are produced by cells in the central nervous system (CNS – primarily the hypothalamus), pancreatic islets (D cells), and epithelial and neuronal areas of the gastrointestinal (GI) tract. There is selective processing of the SS precursor to yield different proportions of the two peptides, which may have different effects in different tissues or on specific functions. SS-14 appears to be the predominant form secreted by neuronal and pancreatic islet cells, whereas SS-28 appears to be the predominant form released by cells in the lower intestine.

The therapeutic applications of SS have been limited because of its very short half-life (1–3 min) in the blood, which calls for continuous intravenous infusion. Pharmacological manipulation of the molecule has produced many analogues that have a longer duration of action and are more potent [87]; one such analogue is octreotide acetate, a cyclic octapeptide that retains the four amino acids of the biologically active region and the cystine bridge of native SS (Fig. 1).

Pharmacology of Octreotide Acetate

The differences between SS and octreotide acetate that reduce the latter's clearance and increase its potency include the substitution of dextroisomers for Trp and the amino-terminal Phe and the reduction of the carboxy-terminal Thr to an alcohol derivative [7]. Octreotide, which is active after subcutaneous administration, has about 45 times the potency of native SS. Peak plasma concentrations of octreotide are observed within 60 min after subcutaneous injection, and its half-life is between 70 and 113 min [25, 48].

Maximal clinical responses to the drug are seen about 2 h after a dose, and the duration of effect varies according to the target tissue considered. For example, the effect is longer on the secretion of thyrotropin (TSH) and vasoactive intestinal peptide (VIP) than on that of growth hormone (GH). Moreover, the suppression of GH levels lasts longer than that of pancreatic hormones, a fact that may account for the selective effects observed in acromegaly [33]. Like SS, octreotide inhibits the nocturnal rise of both GH and TSH and suppresses their release in response to stimuli [25, 33]. Al-

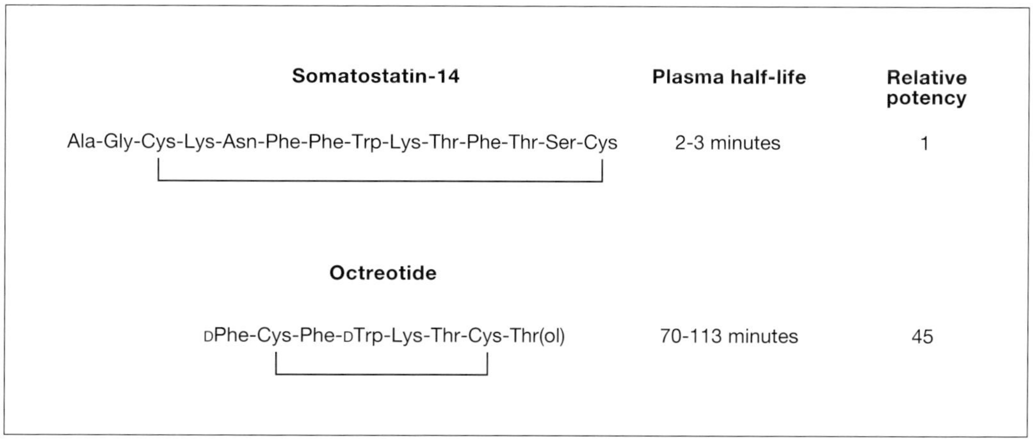

Fig. 1. *Comparison of SS and octreotide acetate*

	Plasma half-life	Relative potency
Somatostatin-14		
Ala-Gly-Cys-Lys-Asn-Phe-Phe-Trp-Lys-Thr-Phe-Thr-Ser-Cys	2-3 minutes	1
Octreotide		
DPhe-Cys-Phe-DTrp-Lys-Thr-Cys-Thr(ol)	70-113 minutes	45

though TSH levels are suppressed, thyroxine concentrations remain in the normal range. Fasting blood glucose levels are unchanged after octreotide injections in nondiabetics, but the response to an oral glucose challenge is impaired.

Octreotide inhibits the postprandial release of various peptides of the gastroenteropancreatic (GEP) system and reduces gallbladder contraction and bile output [48].

The Effects of Somatostatin and Octreotide on the Pituitary

The hypothalamic cells, principally in the anterior periventricular region and the medial division of the paraventricular nucleus, secrete SS directly into the hypophyseal portal system. SS inhibits both GH and TSH secretion from the anterior pituitary [37] (Fig. 2). Conversely, high levels of GH and insulin-like growth factor I (IGF-I) increase hypothalamic SS secretion. In addition, TSH and thyroid hormone are involved in the feedback regulation of SS secretion [77].

SS infusion in normal individuals suppresses the basal and nocturnal secretion of GH and clearly inhibits the GH response to exercise, insulin-stimulated hypoglycemia, and growth-hormone-releasing hormone (GHRH) [37, 75]. Rebound hypersecretion of GH is observed after discontinuation of SS infusion, a fact indicating that the primary effect of SS is on the release, not the synthesis, of GH. Whereas the nocturnal rise in TSH levels and the TSH response to thyrotropin-releasing hormone (TRH) are eliminated by SS infusion [37, 100], stopping the infusion causes no rebound hypersecretion of TSH: unlike its effects on somatotrophs, SS appears to inhibit both the release and the synthesis of TSH by thyrotrophs. SS has no effect on the secretion of prolactin (PRL) and adrenocorticotropin (ACTH). The infusion of SS in normal volunteers did not affect the basal or stimulated secretion of PRL or ACTH [37] or the basal gonadotropin (Gn) levels, but it did inhibit the release of luteinizing hormone (LH) following the administration of Gn-releasing hormone (Gn-RH) [16, 37].

Effects on Pituitary Tumors

GH-Secreting Tumors

In adults, GH hypersecretion from pituitary tumors produces acromegaly [4, 69]. Excess GH stimulates the production of IGF-I in many tissues, and IGF-I levels reflect the integrated value of pulsatile GH secretion and correlate with clinical evidence of active disease such as soft tissue and bony changes. The enlarging tumor may, in addition, cause signs and symptoms of intracranial compression.

The treatment of patients with GH-secreting tumors is aimed at reducing the effects of excess GH and at preventing tumor growth. Specifically, therapy is aimed at reducing mean GH levels in the blood to less than 5 µg/l, which correlates with normal IGF-I blood levels of less than 2 U/ml [5]. Transsphenoidal hypophysectomy, which produces an acute reduction in GH levels and provides significant symptomatic relief, is the treatment of choice. However, about 85 % of patients have macroadenomas, and less than half of these patients are permanently cured by surgery. Conventional supervoltage or proton-beam radiation are effective in reducing GH hypersecretion and preventing tumor growth but require years to achieve maximal effects [38, 40]. Bromocriptine, a dopamine agonist, provides symptomatic relief but is not effective in completely suppressing GH hypersecretion [34]. Additional forms of therapy are needed for patients who are inadequately treated by surgery or irradiation and are resistant to bromocriptine.

Because of its short half-life, natural SS has been little used in the treatment of pituitary tumors. In acromegaly, SS infusions have been shown to suppress GH levels and induce abnormal glucose tolerance profiles [11]. The acute effects of octreotide result in the prompt reduction of GH secretion. Peak height and nadir concentrations of GH in the blood decrease, but there appears to be an increase in the pulse frequency of GH release [82] suggesting that octreotide acts on GH hypersecretion by the tumor but does not inhibit the hypothalamic secretion of GHRH.

In short-term studies, octreotide significantly reduced GH concentrations in about 80 % of patients [19], but reduction to below 5 µg/l was observed less frequently. Significant and sustained reduction in GH concentration by octreotide treatment has usually been observed in patients with pretreatment GH levels below 30 µg/l. The efficacy of octreotide in suppressing GH and IGF-I levels has also been demonstrated in a recent short-term, randomized, placebo-controlled study of 20 acromegalic patients [30]. Whereas in the placebo group 50 %–150 % fluctuations from baseline GH levels were observed over 2 weeks, in the octreotide-treated patients the mean GH concentrations decreased by more than 50 % in eight out of ten patients. Furthermore, significant reductions in IGF-I levels were observed in the group receiving octreotide. More patients noted clinical improvement on octreotide than on placebo.

The response of GH-secreting tumors to octreotide, like that to SS, appears to be dependent on the expression of the SS receptors [41]. In acromegalic patients responding to octreotide with a suppression of GH levels to less than 15 µg/l, SS receptors were clearly demonstrable in the tumors by autoradiography [79]. The expression of SS receptors by various neoplastic endocrine cells has made possible diagnostic imaging using radiolabeled octreotide to locate these tumors [54], and positive imaging appears to predict a favorable clinical response to octreotide therapy.

Many studies have examined the long-term effects of octreotide in the treatment of acromegaly. Most patients had previously undergone surgery or irradiation and were resistant to bromocriptine treatment. Prolonged uninterrupted administration of octreotide (100–1600 µg/day) reduced GH levels [34, 51] and normalized IGF-I levels in about 60 % of patients. The signs and symptoms of acromegaly – headaches, hyperhidrosis, arthralgias, and soft tissue swelling – improved, and reduction of blood pressure and weight have been observed in some patients. Both improve-

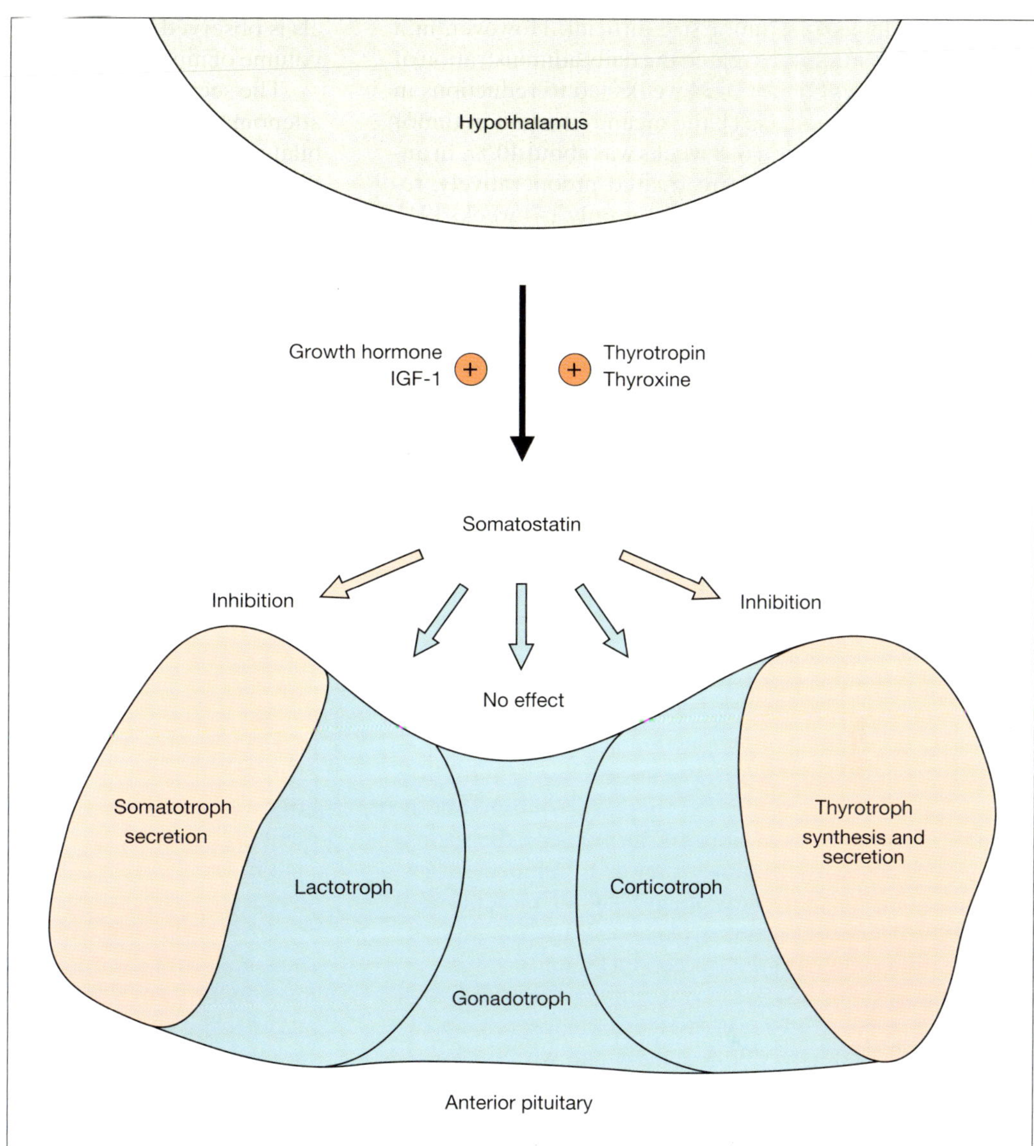

Fig. 2. *Effects of SS on the pituitary*

ment and worsening of blood glucose control have been reported in acromegalic patients with associated diabetes mellitus. One study has reported the reduction of the hyperdynamic cardiac features and the improvement of symptoms of associated heart failure in patients on prolonged octreotide therapy [15].

It is now evident from several studies that the optimum dose of octreotide for the long-term treatment of GH hypersecretion is 300 µg/day [40, 76, 86]. In one large study [86], the optimum dose for mean 12-h GH suppression was 300 µg in 50% of patients, 1500 µg being optimal in only 20% of patients. In another study, normalization of IGF-I levels to less than 2 U/ml was achieved in 43% of patients with 300 µg/day or less [76]; dose increases to 600 or 1500 µg/day increased the responder rate by only 20%.

The optimum regimen for most acromegalic patients on long-term treatment is probably 100 µg octreotide every 6–8 h [40], a regimen that also takes into account the compliance problem associated with multiple daily injections. In general, no additional benefit results from more frequent injections of octreotide or from its being infused.

In a randomized study of 23 acromegalic patients comparing bromocriptine 22.5 mg/day and octreotide 300–600 µg/day for 8 weeks, there were no between-group differences in mean 12-h GH or IGF-I concentrations [86]. However, these levels were normalized on octreotide treatment in twice as many patients as on bromocriptine. Soft tissue swelling, joint stiffness and pain, and hyperhidrosis improved equally in both groups, yet octreotide appeared to be more effective in relieving headaches and restoring vitality. It is still unclear whether combination therapy using both drugs is a worthwhile therapeutic option.

Many studies have described tumor shrinkage with octreotide, as determined by computed tomography (CT) scans [34]. In some instances, the patients' having being previously treated with irradiation or surgery can make the assessment of changes in tumor size difficult. However, in a study of 10 patients not so treated the daily administration of 300–1000 µg octreotide for 3–34 weeks led to reductions in tumor size in all patients [31]; the mean reduction in tumor volume over an average of 6–8 weeks was about 40 %. In another study, five patients were treated preoperatively, receiving only 300 µg octreotide/day for only 1–4 weeks [94]; in only one patient was there evidence of a modest reduction in tumor size. The antitumor effect of octreotide is probably small when compared with its inhibitory effect on GH secretion.

There is one report on the occurrence of a pregnancy during octreotide treatment; exposure of the fetus to the drug during the first 2 months of gestation did not cause malformations or affect its development [55].

TSH-Secreting Tumors

TSH-secreting pituitary tumors are rare causes of goitrous hyperthyroidism [92]. TSH and its free α-subunit are secreted in large quantities and appear not to be regulated by thyroid hormone levels. The identification of patients with these tumors is difficult, and they are usually misdiagnosed as having commoner forms of goitrous hyperthyroidism. Thyroid ablation may resolve the hyperthyroidism, but the resultant hypothyroidism appears to enhance the growth of the TSH-secreting pituitary tumor. Combined surgery and radiotherapy cures less than 50% of patients [92]. Bromocriptine acutely reduces TSH levels in a few patients, but long-term treatment is unsuccessful [83].

In patients with TSH-secreting tumors, SS infusions induce a 20 %–40 % decrease in TSH levels in responders. Octreotide is also effective, both acutely and with prolonged treatment [8, 21]. In doses smaller than 300 µg/day, the analogue normalizes the plasma concentrations of TSH and restores euthyroidism. Reduction in α-subunit levels is also observed.

TSH-secreting tumors appear to be more sensitive to the effects of octreotide than GH-secreting tumors, since both markers of the former are normalized with lower doses (50–150 µg/day) than are required in acromegaly. Unlike with GH-secreting tumors, octreotide appears to inhibit the synthesis as well as the release of TSH, as shown by the gradual return to pretreatment levels of TSH upon discontinuation of octreotide. In one patient, prolonged octreotide treatment resulted in tumor shrinkage and resolution of visual-field impairment [35].

Other Pituitary Tumors

There are few studies of SS or octreotide treatment in patients with non-GH- or non-TSH-secreting tumors. While ineffective in the treatment of pure prolactinomas [52], which fortunately respond well to dopamine agonists, octreotide acutely suppresses both GH and PRL secretion in acromegalic patients with mixed GH/PRL-secreting adenomas. Furthermore, continued reduction of GH and PRL levels is observed on prolonged therapy, and reductions in the volume of mixed tumors have been observed.

The secretion of excess ACTH in patients with pituitary adenomatous development (Nelson's syndrome) following bilateral adrenalectomy has been shown to be suppressed by SS infusions [96] or by octreotide [53]. In one patient, visual-field defects were improved after 6 weeks of treatment despite the absence of tumor shrinkage on CT scan, and continued suppression of ACTH levels was demonstrated over 2 years. Octreotide treatment of patients with pituitary-dependent hypercortisolism resulted in no change in blood ACTH and cortisol levels [53].

No report is available on the use of SS in Gn-secreting or nonsecreting pituitary adenomas. Octreotide therapy failed to suppress Gn hypersecretion in two patients with Gn-secreting tumors [70, 98].

Side Effects of Octreotide

The adverse effects of octreotide therapy are known mainly from studies in patients with pituitary tumors. They include nausea with or without vomiting, pain at the injection site (which can be reduced by warming the solution), diarrhea or obstipation, and abdominal pain [34, 51]. Most of these side effects are mild and transient and only rarely lead to treatment discontinuation.

The side effects of long-term octreotide treatment are similar to the clinical manifestations of the somatostatinoma syndrome, which are diabetes mellitus, cholelithiasis, diarrhea, malabsorption, and weight loss [49]. Although several investigators have reported abnormal glucose tolerance in acromegalic patients receiving octreotide, frank diabetes is rare [38, 68, 90]. Still, the development of abnormal glucose tolerance gives rise to concern about a possible increased risk for cardiovascular disease. Fecal fat loss is an inconsistent finding [67, 85]. Significant fat losses have been reported, but no clinical or biochemical evidence of abnormalities related to fat malabsorption has been seen. In acromegalic patients given octreotide, biliary sludge, cholesterol crystal formation, and microlithiasis have been detected by ultrasound examination of the gallbladder [34, 51, 85]. Large stones have been observed too, as well as acute cholecystitis and cholangitis [40, 74]. Cholelithiasis probably results from octreotide-induced changes in bile composition and the inhibition of pancreatic hormone secretion and gallbladder contraction. More studies are needed to determine the frequency and consequences of biliary changes on ultrasound examination and to evaluate the need for intervention to prevent gallbladder disease [23].

Elevation of hepatic transaminases has been reported [34]. An unexpected but common finding of moderate to severe gastritis has been made in acromegalic patients on long-term octreotide treatment [74], although the patients had not been examined in this respect before the start of octreotide treatment. In a patient with acromegaly and a nonsecreting pancreatic tumor, melena was reported to have developed after 6 months of octreotide treatment and after a later challenge with the drug [86].

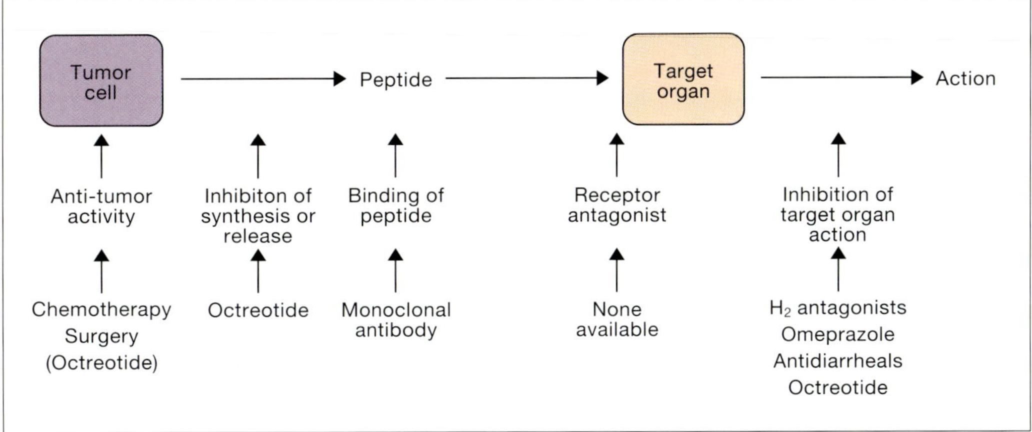

Fig. 3. *Potential sites of action of drugs for neuroendocrine tumors producing clinical syndromes*

The Effects of Somatostatin and Octreotide on Neuroendocrine Tumors of the Gut

Neuroendocrine tumors of the gut (also called GEP neuroendocrine tumors) are usually classified functionally into those that produce mainly serotonin (carcinoids and the carcinoid syndrome), VIP (VIPomas), insulin (insulinomas), glucagon (glucagonomas), gastrin (gastrinomas), SS (somatostatinomas), or growth-hormone-releasing factor/hormone (GHRFomas, GHRHomas) [17, 20, 31, 43, 62, 64]. Such tumors may also release ACTH or humoral hypercalcemic factors [17, 31, 62]. Only rarely do gut tumors produce solely neurotensin (neurotensinomas) or pancreatic polypeptide (PPomas), which give rise to no clinical syndrome and are thus referred to as nonfunctioning tumors in spite of their being neuroendocrine in type by light microscopy [17]. Many neuroendocrine tumors release more than one peptide, e.g., VIP and PP or gastrin and ACTH, and may contain yet other peptides detectable by immunocytochemistry; such tumors are usually classified according to the dominant clinical syndrome that they cause.

Carcinoids are the most common neuroendocrine tumors. They occur principally in the terminal ileum but can be found throughout the intestine and also in the lungs and ovaries [62]. The vast majority of them give rise to no clinical syndrome, but on occasion lung or ileal carcinoids metastasize to the liver and produce the carcinoid syndrome [62], characterized by flushing of the upper half of the body and diarrhea, with or without wheezing, dyspnea, heart failure, mental changes, arthropathy, and even a pellagroid rash [62]. The carcinoid syndrome is always associated with evidence of increased serotonin metabolism, which is most easily revealed by increased urinary amounts of the metabolite 5-hydroxyindoleacetic acid (5-HIAA).

The estimated incidence of the carcinoid syndrome is about 15 new cases per million population per year [62], as against 0.9 for insulinomas, 0.4 for gastrinomas, and less than 0.2 for all other GEP tumors [14]. Any such tumor can occur sporadically as an isolated tumor or be part of familial multiple endocrine neoplasia, type I. All gut neuroendocrine tumors tend to grow slowly, and some patients can live for many years with metastatic disease. Nevertheless, because of the release of biologically active products into the circulation, these tumors can cause severe symptoms even when they are still localized, and death may result from the consequences of peptide overproduction rather than tumor progression. Conversely, the abolition of paraneoplastic symptoms can leave the patient feeling well despite extensive hepatic metastases.

The treatment of neuroendocrine tumors (Fig. 3) has included surgery either to remove the primary tumor completely or to reduce tumor bulk in cases with metastases, whereas for inoperable metastatic tumors either chemotherapy or hepatic arterial embolization to induce the ischemic necrosis of hepatic metastases has been used [2, 14, 17, 31, 43, 63].

In patients with Zollinger-Ellison syndromes (gastrinomas), an alternative approach for symptomatic relief is to block the function of the hydrochloric-acid-producing parietal cells by preventing gastrin from stimulating them; this is done with an antagonist of histamine H_2-receptors or with the H^+, K^+-ATPase inhibitor omeprazole [43, 66]. No comparably effective drugs have been developed for other GEP endocrine tumors, although antidiarrheal drugs may be of limited use in some patients with VIPomas [17, 31].

A third approach is to reduce by pharmacological means the amounts of circulating biologically active products. The administration of natural SS to normal volunteers reduces the basal and stimulated plasma concentrations of many circulating peptides [12, 13, 47, 77, 88], and also inhibits gastric and pancreatic secretion and intestinal absorption and motility by a direct action [77]. SS infusions have been shown to produce biochemical and sometimes symptomatic improvement in patients with the carcinoid syndrome [32, 59], VIPoma [1, 56, 58, 84], insulinoma [1, 9, 10, 18, 22, 26, 57], gastrinoma [1, 12, 28], and glucagonoma [1, 27, 46, 93]. However, natural SS has a half-life in the circulation of about 3 min [91] and is therefore of limited therapeutic utility.

The development of the long-acting SS analogue octreotide has made the third approach practicable. Because neuroendocrine gut tumors are rare, no controlled trials have been done in these indications. In early studies, most patients were initially given 50 µg every 12 h, with subsequent increases to 600 µg/day; if no symptomatic response was achieved, the drug was discontinued. Meanwhile, the dose limit has been raised to 1500 µg/day. Although octreotide has been used mainly for symptom relief, various SS analogues – including octreotide – show antitumor activity in animal models [78, 81]. As will be seen below in the relevant section, the effect of octreotide on tumor size has been studied in a few long-term studies, but the data are still fragmentary.

The Carcinoid Syndrome

There are reports on more than 100 cases treated with octreotide [34, 61] in doses of 50 µg every 12 h to 500 µg every 8 h. In more than 90% of the patients treated, the symptoms (flushing or diarrhea or both) improved during therapy. In addition, urinary 5-HIAA fell in two-thirds of patients for whom the relevant data were given, but in no patient did 5-HIAA become normal. In many patients, the beneficial effects of octreotide persisted for many months without a change in the dose. When symptoms recurred on therapy, increasing the dosage controlled the symptoms in some but not all patients. Two patients have been reported in whom octreotide was useful or life-saving in a carcinoid crisis: one patient responded favorably to a continuous infusion of octreotide 50 µg/h, and the other patient was successfully resuscitated with 100 µg octreotide injected intravenously.

Chemotherapy is not very effective for symptom relief in the carcinoid syndrome; hepatic artery embolization is effective but entails significant risks; and interferon, although promising, has many side effects. Octreotide, which has proved useful for ameliorating the symptoms in most patients, thus represents a significant advance in therapy. It should be readily available whenever there is a risk of precipitating a carcinoid crisis.

Insulinomas

The excessive amounts of insulin released by insulinomas cause hypoglycemia and transient neuropsychiatric disturbances associated with palpitations and sweating. Unlike other islet cell tumors, which – at least by the time they are diagnosed – are usually associated with the presence of hepatic metastases, insulinomas are localized and benign in about 90% of cases and are thus best treated by resection [20]. There are reports of more than 30 patients with insulinomas given octreotide in short-term studies [61], typically 50 µg or more every 12 h. The effects on plasma insulin and glucose were not consistent, and in some patients octreotide made the hypoglycemia worse, possibly because the production of counterregulatory peptides (glucagon and GH) was more suppressed than was that of insulin by the tumor. In most patients with unresectable insulinomas given octreotide for more than 1 month, hypoglycemia was well controlled, despite a return of insulin concentrations to pretreatment levels in some patients. Symptoms recurred with time in about 30% of the patients, and some of them responded to an increase in the dose of octreotide.

Octreotide has proved useful in some patients with metastatic insulinomas or in whom the tumor eluded detection, but the hypoglycemia was sometimes made worse. Besides, diazoxide, an inhibitor of insulin secretion, can be used for symptomatic hypoglycemia. Thus, the role of octreotide in the management of patients with insulinomas remains to be clearly defined.

Gastrinomas

Gastrinomas produce the Zollinger-Ellison syndrome, characterized by elevated plasma concentrations of gastrin and hypersecretion of gastric acid, with resultant peptic ulcerations or diarrhea or both. Of patients with Zollinger-Ellison syndrome, 30%–50% probably have metastatic disease at the time of diagnosis; in 30% of patients without metastatic disease, no primary tumor can be identified [43]. As a consequence, many patients with the Zollinger-Ellison syndrome require long-term symptomatic therapy. Since the clinical manifestations of the syndrome are due to hypersecretion of gastric acid, histamine H_2-receptor antagonists or the H^+, K^+-ATPase antagonist omeprazole, if given in sufficient oral doses, can reduce the acid output to a safe level and render the patients asymptomatic [43, 66].

More than 50 patients with gastrinomas treated with octreotide have been reported [61]. Octreotide lowered the concentrations of plasma gastrin and the secretion of gastric acid at least in part, by a direct action on the stomach. No systematic reliable data are available on the symptomatic responses, because the patients usually also received oral therapy, but some patients were able to reduce their consumption of H_2-receptor antagonists. Since the symptoms of gastrinoma can be effectively and safely controlled by oral medication, there is no definite indication for octreotide in this syndrome.

VIPomas

VIP-producing tumors cause watery diarrhea, hypokalemia, hypochlorhydria, hypophosphatemia, and sometimes hypercalcemia, together with increased plasma concentrations of VIP. Usually the patients have extensive metastatic disease, and virtually all of them become totally resistant to conventional antidiarrheal drugs. Furthermore, chemotherapy is only transiently useful.

Over 20 reports of VIPomas treated with octreotide (100–450 µg/day) have been published [61]. All patients responded initially with a reduction in diarrhea, and the beneficial effects of octreotide were usually evident within 12 h. In some patients, the beneficial effect was striking, but in a few patients it lasted only for a few days and could not be restored by doses as high as 1200 µg/day. Over months, many patients required an increase in the dose to control their symptoms, and others required the addition of steroids. One

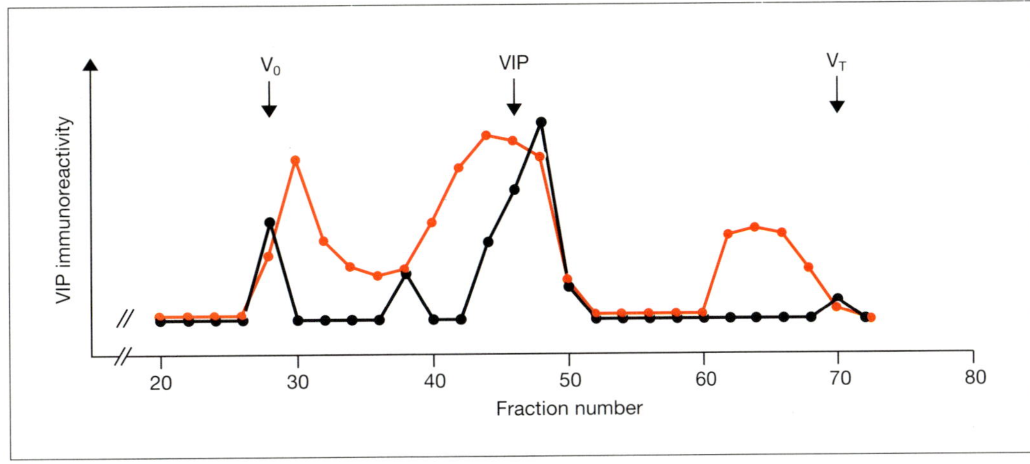

Fig. 4. *G 50 Sephadex chromatography of plasma obtained before treatment* **(black)** *and during octreotide therapy* **(red)** *from a patient with a VIPoma.* V_O, *void volume;* V_T, *total column volume;* **VIP**, *the elution position of synthetic VIP of molecular weight 3055. During octreotide therapy, not only did VIP in plasma fall, but less of the VIP immunoreactivity coeluted with the standard VIP, whereas more eluted as both larger and smaller molecular-weight forms that are probably less bioactive. (To show the differences in molecular weight distribution more clearly, the curves have been drawn in such a way that the largest peaks are of about the same height in both; the peak heights therefore do not reflect the fall in plasma VIP concentration during treatment.) (Data redrawn from [64]*

patient became resistant to octreotide after treatment for more than 1 year. The plasma concentrations of VIP fell during therapy in nearly all patients, but were normalized in only about 30 % of them. However, the symptomatic relief of diarrhea was not always related to changes in the plasma levels of VIP, and in some patients octreotide was effective despite VIP levels that cause secretory diarrhea when VIP is infused into normal volunteers. It may be that the VIP molecule released into the circulation during octreotide treatment is larger and biologically less active than otherwise (Fig. 4); alternatively, octreotide possibly has a direct effect on the gut to reduce fluid secretion.

In patients with VIPomas, octreotide clearly represents a therapeutic advance; since 85 % of patients with VIPomas respond to octreotide, it has to be considered first-line therapy.

Glucagonomas

The glucagonoma syndrome is characterized by an elevated plasma concentration of glucagon, a unique migrating erythematous rash, anorexia, weight loss, mild glucose intolerance, and, on occasions, diarrhea, neuropsychiatric symptoms, and venous thromboses.

More than ten patients with this syndrome, all with metastatic disease, have been given octreotide (100–450µg/day) [61]. When present, the rash resolved in a few days, although two patients were also taking zinc, which is known to ameliorate the glucagonoma-associated rash in some patients. Weight loss, abdominal pain, and diarrhea improved when present, but octreotide had little effect on the diabetes. The plasma concentrations of glucagon fell in nearly all patients but were normalized in only one. In one well-studied patient, the rash resolved with no change in the plasma concentrations of glucagon, amino acids, and zinc (low amino acid and zinc levels are features of the condition that are believed to play a role in the causation of the dermatosis), a fact suggesting that octreotide may have a direct action on the skin. Such a suggestion is consistent with the finding that octreotide can improve psoriasis in some patients. The effectiveness of octreotide was maintained in all patients during long-term treatment, although the plasma concentrations of glucagon returned to pretreatment levels in at least two of them and the dose had to be increased in another two.

Ideally, controlled studies should be performed to establish the efficacy of octreotide in improving the rash, but in the absence of such studies it certainly seems reasonable to use the drug in patients in whom the rash does not respond quickly to simple measures.

GHRHomas

Neuroendocrine tumors producing GHRH stimulate the pituitary to release excessive amounts of GH, causing acromegaly that often proves to be resistant to medical treatment and recurs despite pituitary surgery or radiation or both.

Several patients with GHRHomas in the chest or metastatic GHRHomas in the abdomen have received octreotide [61]. Interestingly, several of the patients initially presented with the gastrinoma syndrome, signs of acromegaly developing only later. Except for those who underwent curative surgical resection of the tumor, these patients received long-term octreotide treatment. All had a good symptomatic response, with reduction in hyperhidrosis in adult patients and arrest of growth and onset of menarche

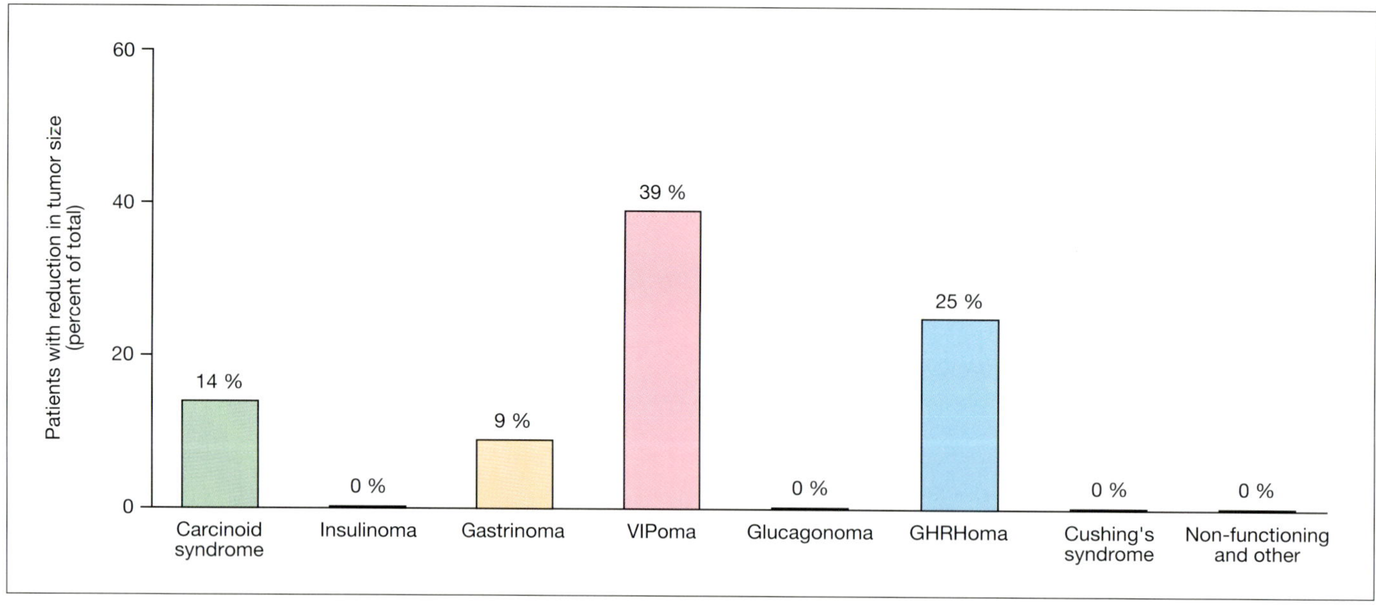

Fig. 5. *Proportions of patients with neuroendocrine tumors of various types in whom reduction in size of metastases was observed during octreotide treatment*

in a prepubescent girl with gigantism. In at least one patient the pituitary became smaller, as measured on CT scans. Octreotide reduced the plasma concentrations of GHRH and GH, but not to the normal range, and thus probably acts in part directly on the pituitary to reduce GH secretion (see the section on GH-secreting pituitary tumors).

Because all these patients, who had proved resistant to other modes of therapy, derived symptomatic relief from octreotide, the drug will continue to be used in patients with GHRHomas who cannot be cured surgically.

Cushing's Syndrome

GEP tumors may produce ACTH and cause ectopic Cushing's syndrome either as an isolated syndrome or, more frequently, in association with the Zollinger-Ellison syndrome [64]. Bronchial carcinoids may also produce Cushing's syndrome. By the time the diagnosis of a GEP ACTH-producing tumor is made, all patients have hepatic metastases; in nearly all cases, the tumors are very aggressive and carry a poor prognosis [31]. There are over ten detailed reports [39, 45, 60, 61] of the effects of octreotide (50 μg every 12 h to 100 μg every 8 h) in patients with neuroendocrine tumors producing ectopic Cushing's syndrome. In most patients the analogue produced clinical improvement in skin changes, blood pressure, diabetes, and hypokalemia. In the cases for which the relevant data were provided, ACTH was normalized, as was urinary free cortisol; in the four cases with additionally elevated plasma gastrin concentrations, these also fell. Octreotide was effective for months in all but one patient. Even if not all cases respond (Maton, unpublished observations), these results suggest that octreotide produces striking improvements in the syndrome of ectopic ACTH secretion.

Tumors that respond to octreotide do not behave like the normal pituitary, whose ACTH secretion is not inhibited by octreotide [3, 42]; the inhibitory effects of octreotide observed in the syndrome of ectopic ACTH production are similar to those on the pituitary of natural SS and octreotide in pathologic states such as Nelson's syndrome [53], adrenal insufficiency [29], or Cushing's disease [53].

Nonfunctioning and Other Rare Neuroendocrine Tumors

A number of patients with neuroendocrine tumors but with no specific endocrine syndrome have been given octreotide [36, 44, 61, 71–73]. The tumors in these patients produced pancreatic polypeptide, neurotensin, calcitonin, glucagon, or no peptide, and all had metastases. The symptoms were nonspecific and included weight loss, abdominal pain, diarrhea, itching, and fatigue. During octreotide treatment, symptomatic improvement was seen in three patients and worsening in one, with one patient's state remaining unchanged. There was no consistent effect on the plasma concentrations of pancreatic polypeptide, glucagon, calcitonin, and neurotensin.

One patient with a tumor producing a parathormone-like peptide with resulting hypercalcemia responded to octreotide for at least 11 months [61]. One patient with a somatostatinoma received two doses of 200 μg octreotide; the fasting plasma concentration of SS was not changed, but the postprandial rise of plasma SS was abolished [24].

In patients with nonfunctioning neuroendocrine tumors there is no indication for octreotide unless it proves to have significant antitumor activity, which seems unlikely (see the next section). In other rare endocrine syndromes produced by GEP tumors, octreotide may be worth trying if there are specific markers by which the drug's effects can be monitored.

Effects of Octreotide on Tumor Size

The effects on tumor size of octreotide 200–1500 μg/day given for 2 months or longer have been studied in over 50 patients with the carcinoid syndrome, 5 with insulinomas, 29 with gastrinomas, 12 with VIPomas, 12 with glucagonomas, 4 with GHRHomas, 4 with Cushing's syndrome, and 5 with nonfunctioning or other neuroendocrine tumors [61]. The metastases increased in size in 24% of patients, stayed unchanged in 62%, and decreased in size in 14% (Fig. 5). there are insufficient data on the natural history of gut neuroendocrine tumors in the general literature and on the rates of tumor growth before therapy in the reports on the use of octreotide to enable definitive statements to be made. The aggregate tumor response rate of 14% is probably an overestimate, since reduction in tumor bulk is more likely to be reported than a negative result. Recently, however, an important German multicenter study has been reported, which examined the effect of octreotide in a well-defined group of patients with metastatic neuroendocrine tumors [95]. All patients were shown by serial CT scans to have actively growing tumors before octreotide therapy (600–1500 μg/day) was carried out according to a strict protocol. No patient had a reduction in tumor size, but tumor size was stabilized initially in 12 of the 17 patients; although there was no control group who received placebo in this study, the data strongly suggest that octreotide had an inhibitory effect on the growth of tumors in the patients treated.

Unresolved Issues

In spite of octreotide's having been in use for several years, questions remain about its mechanism of action, the reasons for differing drug requirements in different patients, and the incidence of – and reasons for – resistance to the drug.

Most neuroendocrine tumors possess SS receptors [50, 80] that account for a direct action of octreotide. Whether the fact that some patients fail to respond or require large doses, whereas others respond readily, is due only to differences in receptor number or affinity is not known. Besides, in some patients octreotide may exert its major effect, not on the tumor and on its peptide production, but on the target organ of this peptide.

How effective is octreotide in long-term studies? Again, few data are available, but one study of seven patients treated for 13–54 months (mean 29 months) found that patients required increasing doses after 4–5 months, became entirely resistant to octreotide after 24±3 months of therapy, and died within 5 months of drug discontinuation [101]. Further long-term data are required to resolve the issue.

References

1. Adrian TE, Barnes AJ, Long RG, et al. (1981) The effect of somatostatin analogs on secretion of growth, pancreatic, and gastrointestinal hormones in man. J Clin Endocrinol Metab 53: 675–681
2. Allison DJ (1982) The nonsurgical management of metastatic endocrine tumor to the liver. In: Wilkins RA, Viamonte M (eds) International radiology. Blackwell, Oxford, pp 191–208
3. Ambrosi B, Bochicchio D, Fadin C, Faglia G (1989) Acute administration of somatostatin or its analog SMS 201≠995 does not influence plasma ACTH levels in patients with corticotrophin hypersecretion (Abstr). J Endocrinol Invest [Suppl 2] 12: 147
4. Barkan AL (1989) Acromegaly diagnosis and therapy. Endocrinol Metab Clin North Am 18: 277–310
5. Barkan AL, Beitins IZ, Kelch RP (1988) Plasma insulin-like growth factor-1/somatomedin-C in acromegaly: correlation with the degree of growth hormone hypersection. J Clin Endocrinol Metab 67: 69–73
6. Barkan AL, LLoyd RV, Chandler WF, Hatfield MK, Gebarski SS, Kelch RP, Beitins IZ (1988) Preoperative treatment of acromegaly with long-acting somatostatin analog SMS 201–995: shrinkage of invasive pituitary macroadenomas and improved surgical remission rate. J Clin Endocrinol Metab 67: 1040–1048
7. Bauer W, Brimer U, Doepfner W, Haller R, Huguenin R, Marbach P, Petcher TJ, Pless J (1982) SMS 201–995: a very potent and selective octapeptide analogue of somatostatin with prolonged action. Life Sci 31: 1133–1140
8. Beck-Peccoz P, Mariotti S, Guillausseau PJ, Medri G, Piscitelli G, Bertoli A, Barbarino A, Rondena M, Chanson P, Pinchera A, Faglia G (1989) Treatment of hyperthyroidism due to inappropriate secretion of thyrotropin with the somatostatin analog SMS 201–995. J Clin Endocrinol Metab 68: 208–214
9. Berchtold P, Berger M, Wiegelmann W, et al. (1976) Serum insulin suppression by somatostatin, diazoxide and phenytoin in insulinoma patients (Abstr 229). 5th International Congress of Endocrinology, Hamburg
10. Berger M, Bordi C, Cuppers HJ, et al. (1983) Functioning and morphologic characterization of human insulinomas. Diabetes 32: 921–931
11. Besser GM, Mortimer CH, Carr D, et al. (1974) Growth hormone release inhibitory hormone in acromegaly. Br Med J 1: 352–355
12. Bloom SR, Mortimer CH, Thorner MO, Hall R, Gomez Pan A, Roy VM, Russel ACG, Coy DH, Kastin AJ, Schalley AV (1974) Inhibition of gastrin and gastric acid secretion by growth hormone release inhibiting hormone. Lancet 2: 1106–1109
13. Boden G, Sivitz MC, Owen OE (1975) Somatostatin suppresses secretin and pancreatic exocrine secretion. Science 190: 163–165

14. Buchanan KD, Johnston CF, O'Hare MM, et al. (1979) Neuroendocrine tumors. A European view. Am J Med [Suppl 6 B] 81: 14–22
15. Chanson P, Timsit J, Masquet C, Warnet A, Guillausseau P-J, Birman P, Harris A, Lubetzki J (1990) Cardiovascular effects of the somatostatin analog octreotide in acromegaly. Ann Intern Med 113: 921–925
16. Chiodera P, Volpi R, d'Amato L, Fatone M, Cigarini C, Fava A, Caiazza A, Rossi G, Coiro V (1986) Inhibition by somatostatin in LHRH-induced LH release in normal menstruating women. Gynecol Obstet Invest 22: 17–21
17. Ch'ng JLC, Polak JM, Bloom SR (1986) Miscellaneous tumors of the pancreas. In: Brooks FP, Lebenthal E, di Magno EP, Scheele GA (eds) The exocrine pancreas: biology, pathobiology and diseases. Raven, New York, pp 763–771
18. Christensen SE, Hansen AP, Lundback K, Orskow H, Seyer-Hansen K (1975) Somatostatin and insulinoma (Letter). Lancet 1: 1426
19. Comi RJ, Gorden P (1987) The response of serum growth hormone levels to the long-acting somatostatin analog SMS 201–995 in acromegaly. J Clin Endocrinol Metab 64: 37–42
20. Comi RJ, Gorden P, Doppman JL, Norton JL (1986) Insulinoma. In: Go VLW, Gardner JD, Brooks FP, Lebenthal E, di Magno EP, Scheele GA (eds) The exocrine pancreas: biology, pathobiology, and diseases. Raven, New York, pp 745–761
21. Comi RJ, Gesundheit N, Murray L, Gorden P, Weintraub B (1987) Response of thyrotropin-secreting pituitary adenomas to a long-acting somatostatin analog. N Engl J Med 317: 12–17
22. Curnow RT, Carey RM, Taylor A, Johanson A, Murad F (1975) Somatostatin inhibition of insulin and gastrin hyper-secretion in pancreatic islet cell carcinoma. N Engl J Med 292: 1385–1386
23. Daughaday WH (1990) Octreotide is effective in acromegaly but often results in cholelithiasis. Ann Intern Med 112: 159–160
24. Davies TME, Bray G, Domin J, et al. (1988) A case of somatostatinoma: responses to food and SMS 201–995 administration. Pancreas 3: 729–733
25. Del Pozo E, Neufeld M, Schlutter K, Tortosa F, Clarenbach P, Bieder E, Wendel L, Neusech E, Marbach P, Cramer H, Kerp L (1986) Endocrine profile of a long-acting somatostatin derivative SMS 201–995. Study in normal volunteers following subcutaneous administration. Acta Endocrinol (Copenh) 111: 433–439
26. Efendic S, Lins PE, Sigurdsson G, Ivemark B, Granberg PO, Luft R (1976) Effect of somatostatin on basal and glucose induced insulin release in five patients with hyperinsulinaemia. Acta Endocrinol (Copenh) 81: 525–529
27. Elsborg L, Glenthoj A (1985) Effect of somatostatin in necrolytic migratory erythema of glucagonoma. Acta Med Scand 218: 245–259
28. Fallucca F, delle Fave G, Giangrande L, del Balzo P, Di Magistris L, Carrutu R (1981) Effect of somatostatin on gastrin, insulin and glucagon secretion in two patients with Zollinger-Ellison syndrome. J Endocrinol Invest 4: 451–453
29. Fehm JL, Voight KH, Lang R, et al. (1986) Somatostatin: a potent inhibitor of ACTH hypersecretion in adrenal insufficiency. Klin Wochenschr 54: 173–175
30. Fredstorp L, Harris A, Haas G, Werner S (1990) Short term treatment of acromegaly with the somatostatin analog octreotide: The first double-blind randomized placebo-controlled study on its effects. J Clin Endocrinol Metab 71: 1189–1194
31. Friesen SR (1982) Tumors of the exocrine pancreas. N Engl J Med 306: 580–590
32. Frolich JC, Bloomgarden ZT, Oates JA, et al. (1978) The carcinoid flush. Provocation by pentagastrin and inhibition by somatostatin. N Engl J Med 299: 1055–1057
33. George S, Hegele RA, Burrow GN (1987) The somatostatin analogue SMS 201–995 in acromegaly: prolonged, preferential suppression of growth hormone but not pancreatic hormones. Clin Invest Med 10: 309–315
34. Gorden P, Comi RJ, Maton PN, Go WVL (1989) Somatostatin and somatostatin analogue (SMS 201–995) in treatment of hormone-secreting tumors of the pituitary and gastrointestinal tract and non-neoplastic diseases of the gut. Ann Intern Med 110: 35–50
35. Guillausseau PJ, Chanson P, Timsit J, Warnet A, Lajeunie E, Duet M, Lubetzki J (1987) Visual improvement with SMS 210–995 in a patient with thyrotropin-secreting pituitary adenoma. N Engl J Med 317: 53–54
36. Guliana JM, Guillausseau PJ, Caron J, Siame-Mouror C, Calmettes C, Modigliani E (1989) Effects of short-term subcutaneous administration of SMS 201–995 on calcitonin plasma levels in patients suffering from medullary thyroid carcinoma. Horm Metab Res 21: 584–586
37. Hall R, Bessor GM, Schally AV, Coy DH, Evered D, Goldie DJ, Kastin AJ, McNeilly AS, Mortimer CH, Phenekos C, Tunbridge WMG, Weightman D (1973) Actions of growth hormone-release inhibitory hormone in healthy men and acromegaly. Lancet 2: 581–584
38. Halse J, Harris AG, Kvistborg A, Kjartansson O, Hanssen E, Smiseth O, Djosland O, Hass G, Jervell J (1990) A randomized study of SMS 201–995 versus bromocriptine treatment in acromegaly: clinical and biochemical effects. J Clin Endocrinol Metab 70: 1254–1261
39. Hearn PR, Reynolds CL, Johansen K, Woodhouse NJY (1988) Lung carcinoid with Cushing's syndrome: control of serum ACTH and cortisol levels using SMS 201–995 (sandostatin). Clin Endocrinol (Oxf) 28: 181–185
40. Ho KY, Weissberger AJ, Marbach P, Lazarus L (1990) Therapeutic efficacy of the somatostatin analog SMS 201–995 (octreotide) in acromegaly. Ann Intern Med 112: 173–181
41. Ikuyama S, Nawata H, Kato KI, Ibayashi H, Nagagaki H (1986) Plasma growth hormone responses to somatostatin (SRIH) and SRIH receptors in pituitary adenomas in acromegalic patients. J Clin Endocrinol Metab 62: 729–733
42. Invitti C, de Martin M, Brunani A, Pecori Giraldi F, Piolini M, Cavagnini F (1989) Effect of a long-acting somatostatin analogue, SMS 201–995, on ACTH secretion in normal subjects and in patients with Cushing's syndrome (Abstr). J Endocrinol Invest [Suppl 1] 12: 201
43. Jensen RT, Doppman JL, Gardner JD (1986) Gastrinoma. In: Go VLW, Gardner JD, Brooks FP, Lebenthal E, di Magno EP, Scheele GA (eds) The exocrine pancreas: biology, pathobiology, and diseases. Raven, New York, pp 727–744
44. Jerkins TW, Sacks H, O'Dorisio T, Solomon S (1986) Long-acting somatostatin analog (SMS 201–995) suppresses elevated pancreatic polypeptide (PP) but not calcitonin (CT) in medullary carcinoma of thyroid (Abstr). Clin Res 34 (1): 196 A
45. Johansen K, Reid K, Woodhouse N (1988) Acute ACTH-lowering effect of SMS 201–995 (sandostatin) in a patient with Cushing's syndrome due to ectopic ACTH-producing lung carcinoid. Saudi Med J 9 (5): 512–514
46. Kahn CR, Bhathena SJ, Recant L, Rivier J (1981) Use of somatostatin and somatostatin analogs in a patient with a glucagonoma. J Clin Endocrinol Metab 53: 543–549
47. Koerker DJ, Ruth W, Chideckel E (1974) Somatostatin: hypothalamic inhibitor of the endocrine pancreas. Science 184: 482–484
48. Kraenzlin ME, Wood SM, Neufeld M, Adrian TE, Bloom SR (1985) Effect of long acting somatostatin-analogue, SMS 201–995, on gut hormone secretion in normal subjects. Experientia 41: 738–740
49. Krejs G, Orci L, Conlon JM, Ravazzola M, Davis GR, Raskin P, Collins SM, McCarthy DM Baetens D, Rubenstein A, Aldor TAM, Unger RH (1979) Somatostatinoma syndrome: biochemical, morphologic, and clinical features. N Engl J Med 301: 285–292
50. Krenning EP, Bakker WH, Breeman WAP, Koper JW, Kooij PPM, Ausema L, Lameris JS, Reubi JC, Lamberts SWJ (1989) Localization of endocrine-related tumors with radio-iodinated analogue of somatostatin. Lancet 1: 242–244
51. Lamberts SWJ (1988) The role of somatostatin in the regulation of anterior pituitary hormone secretion and the use of its analog in the treatment of human pituitary tumors. Endocr Rev 9: 417–436
52. Lamberts SWJ, Zweens M, Klijn JGM, van Vroonhoven CCJ, Stefanko SZ, del Pozo E (1986) The sensitivity of growth hormone and prolactin secretion to the somatostatin analogue SMS 201–995 in patients with prolactinomas and acromegaly. Clin Endocrinol (Oxf) 25: 201–212

53. Lamberts SWJ, Uitterlinden P, Klign JMG (1989) The effect of the long-acting somatostatin analogue SMS 201–995 on ACTH secretion in Nelson's syndrome and Cushing's disease. Acta Endocrinol (Copenh) 120: 760–766
54. Lamberts SWJ, Bakker WH, Reubi J-C, Krenning EP (1990) Somatostatin-receptor imaging in the localization of endocrine tumors. N Engl J Med 323: 1246–1249
55. Landolt AM, Schmid J, Wimpfheimer C, Karlsson ERC, Boerlin V (1989) Successful pregnancy in a previously infertile woman treated with SMS 201–995 for acromegaly. N Engl J Med 320: 671–672
56. Lennon JR, Sircus W, Bloom SR, et al. (1975) Investigation and treatment of a recurrent vipoma. Gut 16: 821–822
57. Lins PE, Efendic S (1979) Responses of patients with insulinomas to stimulators and inhibitors of insulin release that have been linked with cyclic adenosine monophosphate. Diabetes 28: 190–195
58. Long RG, Barnes AJ, Adrian TE, et al. (1979) Suppression of pancreatic endocrine tumor secretion by long-acting somatostatin analogue. Lancet 2: 764–767
59. Long RG, Peters JR, Bloom SR, Brown MR, Vale W, Rivier JE, Grahame-Smith DG (1981) Somatostatin, gastrointestinal peptides and the carcinoid syndrome. Gut 22: 549–553
60. Lorcy Y, Delambre C, Leguerrier AM, Bouyaux M, Deidier A, Allannic H (1988) Traitement de la maladie de Cushing par un analogue de la somatostatine (SMS 201–995): quatres observations. Ann Endocrinol (Paris) 50: 311
61. Maton PN (1989) The use of the long acting somatostatin analogue, octreotide acetate in patients with islet cell tumors. Gastroenterol Clin North Am 18: 897–922
62. Maton PN, Hodgson HJ (1984) Carcinoid tumors and the carcinoid syndrome. In: Bouchier IAD, Allan RN, Hodgson HJF, Keighley MRB (eds) Textbook of gastroenterology. Bailliere-Tindall, London, pp 620–634
63. Maton PN, Camilleri M, Griffin G, Allison DJ, Hodgson HJ, Chadwick VS (1983) Role of hepatic arterial embolization in the carcinoid syndrome. Br Med J 287: 932–935
64. Maton PN, Gardner JD, Jensen RT (1986) The incidence and etiology of Cushing's syndrome in Zollinger-Ellison syndrome. N Engl J Med 315: 1–5
65. Maton PN, O'Dorisio TM, O'Dorisio MS, Malarkey WB, Gower WR Jr, Gardner JD, Jensen RT (1986) Successful therapy of pancreatic cholera with the long-acting somatostatin analogue SMS 201–995: relation between plasma concentrations of drug and clinical and biochemical responses. Scand J Gastroenterol [Suppl 119] 21: 181–186
66. Maton PN, Gardner JD, Jensen RT (1989) Recent advances in the management of gastric acid hypersecretion in patients with Zollinger-Ellison syndrome. Gastroenterol Clin North Am 18: 847–863
67. McGregor AR, Troughton WD, Donald RA, Espiner EA (1990) Effects of the somatostatin analogue SMS 201–995 on faecal fat excretion in acromegaly. Horm Metab Res 22: 55–56
68. McKnight JA, McAnce DR, Crothers JG, Atkinson AB (1989) Changes in glucose tolerance and development of gallstones during high dose treatment with octreotide for acromegaly. Br Med J 299: 604–605
69. Melmed S (1990) Acromegaly. N Engl J Med 322: 966–977
70. Micic D, Popovic V, Sumarac M, Kendereski A, Manojloviz D, Micic J (1989) The effects of bromocriptine, LHRH analog and somatostatin analog on hormone secretion by gonadotropin producing pituitary adenoma. J Endocrinol Invest 12: P 172
71. Modigliani E, Chayvialle JA, Cohen R, Perret G, Guliana JM, Vassy R, Roger P, Siame-Mourot C, Bennet M, Bentata-Pessayre M, Baulieu JL, Charpentier G, Ruszniewsky P, Deidier A, Calmettes C (1988) Effect of a somatostatin analog (SMS 201–995) in perfusion on basal and pentagastrin-stimulated calcitonin levels in medullary thyroid carcinoma. Horm Metab Res 20: 773–775
72. Muratori F, Verga U, di Sacco G, Piolini M, Libroia A (1989) Effetti dell' SMS 201–995 sui livelli sierici di calcitonina in pazienti con carcinoma midollare della tiroide (Abstr). J Endocrinol Invest [Suppl 1] 12: 80
73. Muratori F, Verga U, di Sacco G, Piolini M. Libroia A (1989) SMS 201–995 effects on calcitonin serum levels in patients with medullary thyroid carcinoma. J Endocrinol Invest [Suppl 2] 12: 80
74. Plockinger U, Dienemann D, Qabbe H-J (1990) Gastrointestinal side-effects of octreotide during long-term treatment of acromegaly. J Clin Endocrinol Metab 71: 1658–1662
75. Prange Hansen A, Orskov H, Seyer-Hansen K, Lundbaek K (1973) Some actions of growth hormone release inhibitory factor. Br Med J 3: 523–524
76. Quabbe H-J, Plockinger U (1989) Dose-response study and long term effect of the somatostatin analog octreotide in patients with therapy-resistant acromegaly. J Clin Endocrinol Metab 68–873–881
77. Reichlin S (1983) Somatostatin. N Engl J Med 309: 1495–1500, 1556–1563
78. Reubi JC (1984) Somatostatin analogue inhibits chondrosarcome and insulinoma tumor growth. Acta Endocrinol (Copenh) 109: 108–114
79. Reubi JC, Landolt AM (1989) The growth hormone response to octreotide in acromegaly correlate with adenoma somatostatin receptor status. J Clin Endocrinol Metab 68: 844–850
80. Reubi JC, Hacki WH, Lamberts SWJ (1987) Hormone-producing gastrointestinal tumors contain a high density of somatostatin receptors. J Clin Endocrinol Metab 65: 1127–1134
81. Rodding TW, Schally AV (1984) Inhibition of growth of pancreatic carcinomas in animal models by analogs of hypothalamic hormones. Proc Natl Acad Sci USA 81: 248–252
82. Roelfsema F, De Beer H, Frolich M (1990) The influence of octreotide treatment on pulsatile growth hormone release in acromegaly. Clin Endocrinol (Oxf) 33: 297–306
83. Rubello D, Busnardo B, Girelli ME, Piccolo M (1989) Severe hyperthyroidism due to neoplastic TSH hypersecretion in an old man. J Endocrinol Invest 12: 571–575
84. Ruskone A, Rene E, Chayvialle JA, et al. (1982) Effect of somatostatin on diarrhea and on small intestinal water and electrolyte transport in a patient with pancreatic cholera. Dig Dis Sci 27: 459–466
85. Salmella PI, Juustila H, Pyhtinen J, Jokinen K, Alavaikko M, Ruokonen A (1990) Effective clinical response to long term octreotide treatment with reduced serum concentrations of growth hormone, insulin-like growth factor-I, and the amino-terminal propeptide of type III procollagen in acromegaly. J Clin Endocrinol Metab 70: 1193–1251
86. Sassolas G, Harris AG, James-Deider A, French SMS Acromegaly Study Group (1990) Long term effect of incremental doses of the somatostatin analog SMS 201–995 in 58 acromegalic patients. J Clin Endocrinol Metab 71: 391–397
87. Schally AV (1988) Oncological applications of somatostatin analogues. Cancer Res 48: 6977–6985
88. Schegel W, Rapits S, Harvey RF, Oliver JM, Pfeiffer EF (1977) Inhibition of cholecystokinin-pancreozymin release by somatostatin. Lancet 2: 166–167
89. Shen L-P, Pictet RL, Rutter WJ (1982) Human somatostatin. I. Sequence of the cDNA. Proc Natl Acad Sci USA 79: 4575–4579
90. Shepherd JJ, Senator GB (1986) Regression of liver metastases in patients with gastrin-secreting tumor treated with SMS 201–995. Lancet 2: 574
91. Sheppard M, Shapiro B, Pimstone B, Kronheim S, Berelowitz M, Gregory M (1979) Metabolic clearance and plasma half-disappearance of exogenous somatostatin in man. J Clin Endocrinol Metab 48: 50–53
92. Smallridge RC (1987) Thyrotropin-secreting pituitary tumors. Endocrinol Metab Clin North Am 16 (3): 765–792
93. Sohier J, Jeanmougin M, Lombrail P, Passa P (1980) Rapid improvement of skin lesions in glucagonomas with intravenous somatostatin infusion (Letter). Lancet 1: 40
94. Spinas GA, Zapf J, Landolt AM, Stuckmann G, Froesch ER (1987) Pre-operative treatment of 5 acromegalics with a somatostatin analogue: endocrine and clinical observations. Acta Endocrinol (Copenh) 114: 249–256
95. Trautmann ME, Neuhaus C, Bruns C, Hugens-Penzel M, Schwerk WV, Koop H, Arnold R (1990) Secondary failure of growth inhibi-

tion by SMS 201–995 if accompanied by increased hormone levels without loss of SMS receptors. Digestion [Suppl 1] 46: 115–116
96. Tyrell JB, Lorenzi M, Gerich JE, Forsham PH (1975) Inhibition by somatostatin of ACTH secretion in Nelson's syndrome. J Clin Endocrinol Metab 40: 1125–1127
97. Veber DF, Freidinger RM, Perlow DS, Paleveda WJ, Holly FW, Strachan RG, Nutt RF, Arison BH, Homnick C, Randall WC, Glitzer MS, Saperstein R, Hirschmann R (1981) A potent cyclic hexapeptide analogue of somatostatin. Nature 292: 55–58
98. Warnet A, Duet M, Roche D, Chanson P, Timsit J, Seret D, Vincens M, Guillausseau PJ, Lubetzki J (1987) Un nouveau cas feminin d'adénome gonadotrope: reponse a LRH, a TRH, à un agoniste de LRH (buséréline), à la bromocriptine et à la sandostatine. Ann Endocrinol (Paris) 48: A 224
99. Wass JAH (1990) Somatostatin. In: De Groot L (ed) Endocrinology, vol 1. Saunders, New York, pp 152–162
100. Weeke J, Prange Hansen A, Lundbaek K (1975) Inhibition by somatostatin of basal levels of serum thyrotropin (TSH) in normal man. J Clin Endocrinol Metab 41: 168–171
101. Wynick D, Williams SJ, Bloom SR (1986) Resistance to long-term treatment of pancreatic endocrine tumors with the somatostatin analogue octreotide (SMS 201–995) (Abstr). Regul Pept 22: 439

Basic and Clinical Aspects of Neuroscience
Editors.: E. Flückiger, E. E. Müller, M. O. Thorner

Volume 3

The Role of Brain Dopamine

With contributions by P. Riederer, E. Sofic, C. Konradi, J. Kornhuber, H. Beckmann, M. Dietl, G. Moll, G. Hebenstreit, M. O. Thorner, M. L. Vance, J. C. Stoof, F. J. H. Tilders, T. J. Petcher

Cover illustration by J. Haley

1989. IX, 55 pp. 29 figs. 5 tabs. Softcover. ISBN 3-540-50040-5

As in other volumes in the series, this volume conveys up-to-date knowledge in a clear and straightforward manner. It begins with a survey of the neurobiological functions of the brain, with the emphasis on Parkinson's disease. This is followed by a presentation of the role of dopamine in the regulation of human anterior pituitary function. The final two chapters concentrate on the dopamine receptors: first, the binding sites are characterized and the biochemical and physiological consequences of dopamine-receptor stimulation are discussed, and finally, there is a report on the topology of a dopamine-receptor model that can account comprehensively for agonists and antagonists.

R. Nieuwenhuys, University of Nijmegen;
J. Voogd, University of Rotterdam;
C. van Huijzen, University of Nijmegen, The Netherlands

The Human Central Nervous System

A Synopsis and Atlas

3rd rev. ed. 1988. XII, 437 pp. 217 figs. Softcover ISBN 3-540-13441-7

Contents: Introduction. – Gross Anatomy. – Vessels and Meninges. – Brain Slices. – Microscopical Sections. – Functional Systems. – References. – Subject Index.

This book has become a classic in its field. Its main feature is still the outstanding illustrative material: 217 halftones and line drawings that in themselves justify the term "classic". Yet the text of this third edition has been thoroughly revised and extended to cover the most recent concepts and data. For example, it contains an entirely new section on the cerebrovascular system and the meninges.

Also, more attention has been placed on the functional significance of the structures that are discussed and depicted, and numerous correlations with neuro-pathology and clinical neurology have been indicated.

This new edition, just as the previous ones, provides a straightforward, clear and reliable guide to the structural and functional organization of the human central nervous system. Designed primarily for students of medicine and psychology, the book will also be informative for practising neurobiologists. Moreover, it will prove useful as an aide mémoire for specialists in the neurological sciences.

Distribution rights for Japan: Igaku Shoin, Tokyo